Greening the Firm

The Politics of Corporate Environmentalism

Aseem Prakash

The George Washington University

CAMBRIDGE
UNIVERSITY PRESS

PUBLISHED BY THE PRESS SYNDICATE OF THE UNIVERSITY OF CAMBRIDGE
The Pitt Building, Trumpington Street, Cambridge CB2 1RP, United Kingdom

CAMBRIDGE UNIVERSITY PRESS
The Edinburgh Building, Cambridge CB2 2RU, UK http://www.cup.cam.ac.uk
40 West 20th Street, New York, NY 10011-4211, USA http://www.cup.org
10 Stamford Road, Oakleigh, Melbourne 3166, Australia

First published 2000

Printed in the United Kingdom at the University Press, Cambridge

Typeset in Plantin 10/12 pt in QuarkXPress™ [SE]

A catalogue record for this book is available from the British Library

Library of Congress Cataloguing in Publication data

Prakash, Aseem.
Greening the firm: the politics of corporate environmentalism /
Aseem Prakash.
    p.   cm.
Includes bibliographical references and index.
ISBN 0 521 66249 4 (hardbound)
1. Environmental policy Case studies.   2. Industrial management –
Environmental aspects Case studies.   3. International business
enterprises – Environmental aspects Case studies.   4. Baxter
International Inc.   5. Eli Lilly and Company.   I. Title.
GE170.P69   2000
658.4'08–dc21   99-33723   CIP

ISBN 0 521 66249 4 hardback
ISBN 0 521 66487 X paperback

To Papa and Mummy

Greening the Firm

Over the last two decades environmental issues have become important in public and business policy. This book asks why firms sometimes voluntarily adopt environmental policies which go beyond legal requirements. It employs a new-institutionalist perspective, and argues that existing explanations, especially from neoclassical economics, concentrate on external factors at the expense of internal dynamics. Prakash argues that "beyond-compliance" policies are due to two types of intra-firm processes, which he describes as power-based and leadership-based. His argument is supported by analysis of ten cases within two firms – Baxter International Inc. and Eli Lilly and Company – including interviews with managers, and access to meetings and documents. This book therefore examines the internal working of firms' environmental policy in a theoretically rigorous way, providing a significant contribution to the theory of the firm. It will be valuable for students of business and environmental studies, as well as political economy and public policy.

ASEEM PRAKASH is Assistant Professor, Department of Stategic Management and Public Policy at the School of Business and Public Management, Department of Political Science, and The Elliott School of International Affairs, The George Washington University.

Contents

List of tables

List of appendixes

Preface

This project has a long history. After completing my MBA from the Indian Institute of Management, Ahmedabad, I joined Procter and Gamble's marketing department in 1989. I was a great believer in the dominant paradigm that is often taught at business schools: firms maximize profits, those that do not are punished by the market, and managers can objectively pursue profit maximization by employing tools of investment appraisal. So, armed with technocratic knowledge and naive enthusiasm, I joined Procter and Gamble that was (and continues to be) a well-regarded and profitable company. One of the things that really struck me was the role of organizational dynamics in affecting a firm's tactical and strategic decisions. The neoclassical economic theory and the various sophisticated financial and marketing techniques that I had learnt at my alma mater did not seem to have the desired relevance. I had always thought that only the functioning of governmental bureaucracies was impacted by organizational politics. How wrong I was! As I exchanged notes with my fellow MBAs across firms and functional areas, I realized that they too were having similar experiences. Broad structural factors external to firms indeed outlined a framework within which firms made decisions. Firms also seemed to pursue a loosely defined objective of "high" levels of "profits." However, internal politics – inter-personal, inter-departmental, etc. – was important in shaping outcomes. Many projects that were pursued were clearly wrong and many "sensible" policies were not adopted. The strategies and power of key individuals mattered in shaping organizational outcomes.

By 1992, I was ready for a change. Also, Anil Gupta, Indian Institute of Management, Ahmedabad, whom I admire very much, had been encouraging me to pursue a career in academia. In 1993, I enrolled in the Joint Ph.D. program co-sponsored by the School of Public and Environmental Affairs and the Department of Political Science at Indiana University, Bloomington. I planned to focus on environmental issues and it seemed that I had an opportunity to leverage my experience of the corporate world for this task. I, therefore, decided to examine how firms actually make environmental policies. Since most large firms in the US are voluntarily adopting programs that are more stringent than the requirements of the extant laws, I decided to focus on firm-level response to "beyond-compliance" policies. Thus, my research project seemed to have important business strategy and public policy implications.

Beyond-compliance initiatives are also interesting to study because they provide a political space for discursive struggles among managers on the benefits and costs of various environmental programs. As a consequence of such struggles, profits no longer remain an objective and abstract concept that can be technocratically determined. They become a subjective concept that are influenced by varying and changing managerial interpretations of future benefits and costs. As I discovered during my field work, many beyond-compliance projects, even the ones costing hundreds of millions of dollars, were not subjected to formal investment appraisal procedures. Organizational dynamics played an important role in deciding whether or not such projects were adopted. I could discern various kinds of organizational dynamics, the conditions under which they were activated, and their impact on the policymaking processes and on the final policy outcomes. Thus, "unpacking" firms to study environmental policymaking processes provided interesting theoretical and policy-relevant insights.

In summary, the contributions of this study are fourfold. First, it highlights the inadequacy of the neoclassical theory, the dominant paradigm, in explaining why firms selectively adopt beyond-compliance policies. Second, at a broad level, it argues that "agents" (or managers) have some (not complete) autonomy in pursuing beyond-compliance policies; external "structures" alone cannot provide fully specified explanations. Organizational dynamics play an important role. Third, it focuses on the important role of power-based and leadership-based processes in shaping the policies of firms. It argues for "bringing back leadership" in the study of political economy. Further, the book integrates insights from sociological institutional theory and stakeholder theory (that focus on pressures external to firms) with leadership-based and power-based theories of the new-institutionalist tradition. Finally, since the conclusions of this book are generalizable to other issue areas where firms adopt beyond-compliance policies, it outlines important questions for future research.

At Indiana University, I had the privilege of working with the leading scholars of environmental policy, institutional analysis, and political economy. Elinor Ostrom, my mentor, encouraged me to embark on this multi-disciplinary project. As the dissertation committee co-chair and the first reader, she constructively challenged me throughout the process. Lin also provided me with valuable office space at the Workshop in Political Theory and Policy Analysis that enabled me to work in the Workshop's intellectually rich environment and interact with scholars from all over the world. Jeffrey Hart, also my committee co-chair, gave me very useful input on the dissertation and then continually encouraged me to revise and refine it into a book manuscript. Mike McGinnis contributed significantly in shaping this project as well as influencing my overall intellectual agenda. Rosemary O'Leary supported this project from the very outset and introduced me to the idea of examining environmental policies of Baxter International and Eli Lilly. Kerry Krutilla helped me to understand the political economy of environmental issues. Vincent Ostrom has been a source of inspiration and continued intellectual support.

Internal workings of firms on environmental policies are difficult to examine because firms are often unwilling to share information on this subject. I am indebted to Dean James Barnes (School of Public and Environmental Affairs) who wrote to Baxter and Lilly on my behalf, thus paving the way for my empirical work. Managers at Baxter and Lilly were first intrigued by the objectives of my research. This was not surprising since this was the first occasion that an academic researcher was closely scrutinizing them. Eventually, they were very supportive of the project. They candidly discussed policy dynamics of various environmental programs. They made themselves available for multiple interviews and follow-up questions. I had opportunities to attend their environmental policy meetings and to examine their internal documents. However, the interpretations and analyses of the events pertaining to environmental policymaking are entirely mine. Bill Blackburn (Baxter) and Daniel Carmichael (Lilly) enabled me to "step inside" their respective organizations. Don Brannon (Lilly) was extremely helpful in ensuring access to appropriate managers as well as documents. Bill Blackburn, Ron Meissen (Baxter) Don Brannon, and Joan Heinz (Lilly) read my dissertation and provided useful input. I would also like to acknowledge help and support from the following: Rob Currie, Mike Cycyota, Peter Etienne, Munira Meharauli, and Verie Sandborg (all from Baxter); Jonathan Babcock, Catherine Ehlhardt, Bert Gorman, Rick Lattimer, Ronald Pitzer, James Vangeloff, and John Wilkins (all from Lilly).

This book is a significantly revised version of my doctoral dissertation. John Haslam, Cambridge University Press, was very supportive and helpful throughout the review process. The two anonymous referees provided excellent and constructive comments that brought greater clarity to my arguments. My friends and colleagues at Indiana University and George Washington University patiently read and reread numerous drafts and gave me invaluable substantive and editorial comments. These are:

Arun Agrawal, Trevor Brown, Brenda Bushouse, George Candler, Yu-che Chen, Nives Dolsak, Ray Eliason, Dan Kane, Tom Koontz, Matthew Potoski, Susan Sell, and Cynthia Yaudes. Jun-ho Choi and Jennifer Baka provided valuable research and editorial support.

This project involved making numerous visits to Indianapolis (Lilly) and Chicago (Baxter). To defray the costs of fieldwork and other expenses, I received funding from the Center for International Business Education and Research, Kelley School of Business, Indiana University, and the College of Arts and Sciences, Indiana University.

Finally, my family supported me emotionally, intellectually, and financially. I have been fortunate to have wonderful parents who have supported all my ventures including the decision to abandon the lucrative world of Procter and Gamble for a career in academia. I dedicate this book to them. My brother, Anshu, and my sister, Charu, have been a source of great support and affection. They have permitted me to enjoy all the benefits of being the youngest in the family. Nives Dolšak stood by me during my stay at Indiana University and also conceded to do so in the future by marrying me.

1   Greening the firm: an introduction

Though environmental problems have challenged humankind since time immemorial, policy scientists have given serious attention to environmental issues only since the 1960s. A series of industrial accidents and media events such as the publication of Rachel Carson's *Silent Spring* (1962) highlighted the environmental consequences of unfettered industrialization. Responding to public concerns, from the 1970s onwards, the United States Congress has enacted a series of laws stipulating environmental standards and technologies for firms. These policies were often backed by zealous monitoring and enforcement. In the 1980s the policy community and the regulatees began articulating their dissatisfaction with the inefficiencies of command and control policies, specifically questioning the capacities of governmental agencies to implement detailed regulations.

Since the late 1980s, particularly after the Rio Summit of 1992, policymakers appear to have accepted that governmental coercion alone will not be sufficient in forcing firms to adopt environmentally sustainable policies; "right incentives" must be provided (Hahn and Nell 1982; Lee and Misiolek 1986; Baumol and Oates 1988; Oates, Portney, and McGartland 1989; Atkinson and Tietenberg 1991; Tietenberg 1992). More recently, policymakers are beginning to play down their adversarial role, and are highlighting the potential gains of collaborating with firms in developing and implementing environmental policies. Further, as opposed to a reluctance in implementing environmental laws, firms are increasingly inclined to adopt "beyond-compliance" environmental policies, the ones that are more stringent than the requirements of the extant laws and regulations.

Beyond-compliance initiatives could be designed and implemented by regulators, industry associations, or individual firms. For example, in recent years, the United States Environmental Protection Agency (EPA) has launched voluntary beyond-compliance programs such as Green Lights, Project XL, and 33/50. These initiatives are win–win–win for the regulators, firms, and citizens. Regulators are able to implement their mandates to enforce environmental laws at low costs. This is particularly welcome in an era of declining budgets for many governmental programs and of calling for "reinventing government"(Osborne and Gaebler 1992). Citizens enjoy cleaner air and purer water without an increased tax burden. Firms enjoy greater operational flexibility in designing and implementing environmental programs that the command and control era denied to them. Their relationship with regulators also becomes less adversarial.

Regardless of whether regulators view firms as adversaries or as potential partners, as reluctant implementers of extant laws or as enthusiastic participants in beyond-compliance programs, environmental policy scientists have implicitly treated firms as unitary actors with similar responses to external incentives (notable exceptions include Fischer and Schot 1992; Gable 1994; Bunge, Cohen-Rosenthal, and Ruiz-Quintanilla 1995). As a result, there is an inadequate understanding of the internal processes that lead firms to adopt or not adopt various kinds of environmental policies, especially the beyond-compliance ones. In other policy areas and disciplines, however, firms have been "unpacked" and their internal processes extensively studied (March and Simon 1958; Baumol 1959; Cyert and March 1963; Marris 1964; Williamson 1964; Katz and Kahn 1966; Thompson 1973; also, Allison 1971). There is also a well-established literature on the impact of external factors on intra-firm dynamics (Cyert and March 1963; Pfeffer and Salanick 1978; DiMaggio and Powell 1983; Tolbert 1985; Oliver 1991).

In contrast to the existing environmental policy literature, this book examines the processes of environmental policymaking within firms. The *theoretical question* I address is: why do firms *selectively* adopt beyond-compliance environmental policies? Selective adoption implies that a given firm adopts only some but not all policies with similar characteristics, or different firms within the same industry respond differently to a given policy. This study argues that the existing explanations that focus exclusively on factors external to firms and that treat firms as unitary actors are under-specified to answer this question. An examination of intra-firm dynamics is also required. Though factors external to firms create incentives and expectations for managers, intra-firm politics influences how managers perceive and interpret external pressures and act upon them. My *policy question* therefore is: why and how do external factors aid or thwart supporters of beyond-compliance policies to persuade their firms to adopt these policies?

To examine these questions, I explore the following issues. How do managers make decisions on environmental policies? What are the decision criteria? Do managers have different preferences on environmental policies and, if so, do such differences impact policy adoption? Are beyond-compliance policies adopted only if they are projected to deliver adequate levels of quantifiable profits? How are non-quantifiable benefits brought into the equation? Since answers to these questions vary within and across firms, the book investigates internal processes and inter-managerial interactions on environmental policymaking.

Beyond-compliance: an overview

Beyond-compliance is different from over-compliance. In the latter, firms seek to comply with the law but due to technological indivisibilities, deliver more than the legal requirement. Also, adopting uniform technologies across facilities that face varying environmental regulations results in over-compliance (Oates, Portney, and McGartland 1989). In contrast, beyond-compliance policies specifically propose to exceed the requirements of extant laws. They may involve modifying physical aspects of value-addition processes or adopting new management systems.

The profit-maximizing view of the firm predicts that firms will adopt policies that can be demonstrated, *ex ante*, to meet or exceed firms' profit criteria. Thus, from a managerial perspective, environmental policies can be classified along two attributes: (1) whether they meet or exceed the *ex ante* profit criteria as stipulated in capital budgeting or some other established investment appraisal procedure; (2) whether they are required by law or they are beyond-compliance. Based on these attributes, four modal policy types can be identified: Type 1 (beyond-compliance and meet or exceed the profit criteria), Type 2 (beyond-compliance but cannot or do not meet the profit criteria), Type 3 (required by law and meet or exceed the profit criteria) and Type 4 (required by law but cannot or do not meet the profit criteria). This discussion is summarized in table 1.1.

Since Type 3 and Type 4 policies are required by law, firms are expected to adopt them. This is especially true for industrialized countries where environmental laws are perceived by managers as being strictly enforced and penalties for non-compliance are significant. Consequently, most firms are not expected to systematically violate environmental laws. This book, therefore, does not focus on these policies.

Type 1 policies, though not required by law, are consistent with the profit-maximizing model of a firm since they meet the *ex ante* profit criteria. For example, scholars suggest that firms can increase profits by voluntarily reducing pollution (Porter 1991; Porter and van der Linde 1995; Shrivastava 1995; Hart 1995; Russo and Fouts 1997; for a critique, see Walley and Whitehead 1994; Newton and Harte 1996). Such policies enable firms to capture the "low-hanging fruit." It is also suggested that

Table 1.1. *Categories of environmental policies*

Impact on quantifiable profits	Impact on Compliance	
	Ensure compliance	Result in beyond-compliance
Established procedures to assess profitability are employed and the policy meets or exceeds their criteria	Type 4 profitable policies that are required by law; are implemented with low inter-manager conflict	Type 1 policies that involve profitable organizational changes with low inter-manager conflicts
Either established procedures to assess profitability cannot be employed, or if they can be, then they were not employed	Type 3 policies that are required by law; are implemented with low inter-manager conflict if there is stringent punishment for non-compliance and effective monitoring	Type 2 policies that involve inter-manager conflicts

such policies enable firms with greater consumer contact to compete on environmental quality and charge a premium (Arora and Cason 1996). Based on these arguments, these policies seem win–win for virtually every constituent. Of course, due to inertia or lack of knowledge about profit opportunities, firms may be slow to adopt them. Nevertheless, serious opposition within firms to such policies is not expected, and, consequently, this book does not examine them.

In contrast to Type 1, 3, and 4 policies, managers are expected to differ on the economic usefulness of Type 2 policies. This book, therefore, exclusively focuses on these policies. Literature identifies multiple motivations for firms to adopt Type 2 policies. The first category of explanations identifies strategic reasons geared towards potential long-term economic benefits. Firms could preempt and/or shape environmental regulations if they themselves adopt such policies (Fri 1992; Khanna and Damon 1999) and reap first-mover advantages (Nehrt 1998; Porter and van der Linde 1995; for a critique, see Palmer, Oates, and Portney 1995; Rugman and Verbeke 2000). Similarly, technologically advanced firms could raise the cost of entry for their rivals – the assumption being that higher standards will lead to stringent regulations (Barrett 1991; Maloney and McCormick 1982; Salop and Scheffman 1983).

The second set of explanations – sociological institutional theory and stakeholder theory – focus on non-profit objectives of firms that may or may not impact their long-term profit objectives. The institutional theory focuses on the impact of external institutions on the policies of firms

(Scott 1987; Zucker 1988; Oliver 1991; Meyer and Scott, 1992; Hoffman 1997). In contrast to neoclassical economics that privileges two institutions – markets and governments – institutional theory takes into account other social institutions as well. Questioning the atomistic accounts of organizational policymaking, it suggests that firms are not profit maximizers; their policies reflect external pressures for legitimacy. Of course, different institutions have varying capacities to influence firms. This theory would predict that firms adopt Type 2 policies in response to pressures from key external institutions and *managers would have little autonomy in this regard* (Hoffman 1997: 6).[1]

Neoclassical economics views the social objective of business is to maximize shareholders' wealth (Friedman 1970). In contrast, stakeholder theory suggests that firms should (and sometimes do) design policies taking into account the preferences of multiple stakeholders – stakeholders being "any group or individual who can affect or is affected by the achievement of the organization's objectives" (Freeman 1984: 46; Donaldson and Preston 1995; Clarkson 1995). Similarly, the literature on corporate social performance (CSP), responsibility, and responsiveness argues that firms have societal responsibilities other than the pursuit of shareholder wealth maximization (Preston 1975). CSP policies are adopted because they are the "right things to do." Firms could be reactive, defensive, accommodative, and proactive in dealing with them (Wartick and Cochran 1985; Carroll 1995; for a critique see, Wood 1991). It could be argued that since Type 2 policies represent proactive CSP, they are adopted by firms.[2] Of course, different stakeholders and institutions have different expectations; sometimes expectations may even be in conflict (Wood and Jones 1995). Thus, it is critical to examine how managers interpret these expectations and employ them to push their agendas on Type 2 policies.

Though institutional theory and stakeholder theory correctly identify non-profit and long-term (potential) profit reasons for adopting Type 2 policies, they inadequately explain *variations* in response – why do firms *selectively* adopt them? For example, why does firm X consider Policy A

---

[1] Oliver (1991) acknowledges that "agents" may have autonomy even in an institutionalist perspective.

[2] Scholars have examined whether CSP policies positively impact firms' financial performance (Ackerman 1975; Preston and Post 1975; Preston and Sapienza 1990; Jones 1995) and adopting a stakeholder approach furthers firms' economic performance (Cochran and Wood 1984; Barton, Hill, and Sundaram 1989; Kotter and Heskett 1992). These studies have been criticized on theoretical and methodological grounds. As a result, these literatures are inconclusive on the impact of adoption of CSP policies and/or stakeholder approach on firms' economic performance (Wood and Jones 1995; Griffin and Mahon 1997).

but not Policy B as "the right thing to do" although both policies have similar characteristics? Or, why does firm X but not firm Y believe that adopting policy A is the "right thing to do"?

This book draws insights from institutional theory and stakeholder theory and relates them to dynamics within firms. The point of departure is that I do not view managers as passive recipients of external pressures. Since "agents" have autonomy in the realm of Type 2 policies, explanations focusing on external "structures" only are under-specified (Child 1972; Granovetter 1985; Ostrom 1990). Further, managers do not have homogeneous preferences on Type 2 policies. The book focuses on the role of key managers in generating consensus or, if faced with opposition, lobbying the top management to mandate policy adoption. While not denying the importance of external factors, I highlight that in the context of Type 2 policies, managers have autonomy to interpret the impact of external pressures on the long term profit and non-profit objectives. Hence, intra-firm politics is important in explaining variations in adoption within and across firms.

"Unpacking" the firm

To understand internal policy processes, an explication of the notion of a firm is imperative. Neoclassical economic theory treats firms as unitary actors seeking to maximize profits (Hirshleifer 1988). The book interprets its broad message as that firms adopt only those policies and projects that can be demonstrated *ex ante* as potentially profitable. Project appraisal is a technical process and there is a shared understanding among managers about the legitimacy of established appraisal procedures, particularly capital budgeting. This procedure requires estimating future benefits and costs and discounting them with an appropriate discount rate. If a project meets or exceeds a given rate of return, it is deemed potentially profitable. Consequently, capital budgeting ensures that managers examine investment decisions objectively with a focus on maximizing shareholders' wealth. This is an important safeguard for shareholders who often have little say in the running of firms, and are therefore vulnerable to "agency abuses" by managers (Berle and Means 1932; Fama 1980). Further, since maximizing a firm's measurable profits is the primary objective for all managers, policy processes would be consensual. This is not to say that managers have identical preferences on environmental policies. Most likely they do not. However, different managerial preferences are not predicted to play out in the policymaking process because there is consensus that a policy should meet or exceed the profitability criteria.

Capital budgeting is appropriate to assess profitability of projects that involve up-front capital expenditures and generate future cash flows. Since Type 2 policies may not involve capital expenditures, capital budgeting is inappropriate for assessing their profitability. In general, it is difficult to assess the impact on profits of policies that focus on establishing management systems, and hence do not generate revenue or decrease quantifiable costs. To assess the profit impact of Type 2 policies, managers employ subjective methods. Such projects are justified by some managers by arguments such as "they are good for the firm in the long run" and "they are important for keeping the EPA in good humor."

Further, some policies involving significant up-front capital expenditures may be adopted without being subjected to capital budgeting. This suggests that established procedures are not applied consistently and policymaking within firms involves a complex mix of factors. Intra-firm processes, inter-manager interactions, and managerial perceptions of external factors are important in influencing whether or not a Type 2 policy is adopted. Project appraisal is not a technical process only; organizational politics also plays an important role in influencing managerial perceptions of the desirability of a project.

The neoclassical notion of a firm is useful in predicting market outcomes in highly competitive markets or when policies are required by laws that are strictly enforced. It is not helpful in explaining why firms selectively adopt Type 2 policies. For this we need to examine the internal processes of firms. Treating firms as composites consisting of many managers, this book employs a new-institutionalist perspective. Further, it assumes that while maximizing quantifiable profits is often the preeminent goal of most managers, it is not the only goal. Managers also differ in their subjective assessments of the long-term profit impacts of policies. I classify managers into two categories: (1) policy supporters favoring the adoption of beyond-compliance policies whose profit impact is not quantified; and (2) policy skeptics who oppose such policies. There is a third category as well: policy neutral. Since they do not significantly impact policy dynamics, the book does not focus on them.

Within a new-institutionalist perspective, three broad theories of firms can be identified: transaction cost, power-based, and leadership-based. Transaction cost theorists examine an important question that is not adequately addressed by neoclassical economics: why do firms arise at all; alternatively, why and how do managers arrive at "make or buy" decisions? Following Coase (1937), transaction cost theorists view firms as institutions designed to economize on transaction costs by allocating resources through hierarchical fiats and not market mechanisms. Williamson (1975, 1985) focuses on how firms evaluate "make or buy" decisions, and suggests that these decisions reflect managers' desire to minimize transaction costs given asset specificity, bounded rationality, and a potential for labors' opportunism. Transaction cost theories, however, do not specifically address my research question: why do firms selectively adopt beyond-compliance policies? Consequently, the book employs power-based and leadership-based explanations only to examine intra-firm dynamics. Importantly, both theories focus on the crucial role of organizational politics – especially the preferences, strategies, and endowments of key managers – in shaping policy outcomes.

There is extensive literature suggesting that managers are "boundedly rational," often have heterogeneous preferences, and are organized as coalitions that seek different policy objectives (Cyert and March 1963; Simon 1957). Since boundedly rational managers make decisions under uncertainty, decision making is often influenced by inter-managerial interactions. Employing these insights, this book suggests that beyond-compliance policies provide political space for "discursive struggles" (Hajer 1995) within firms on their long-term profit and non-profit impact. If such policies are adopted, it is by two kinds of processes: (1) power based, where policy supporters, in face of opposition from policy skeptics, "capture" the top management and have it mandate the adoption of such policies; (2) leadership based, where policy supporters succeed in inducing consensus, convincing policy skeptics and policy neutrals of the long-term benefits of such policies. It is important to differentiate power-based from leadership-based processes since they arise under different conditions and lead to different types of outcomes. In both processes, managers invoke the external environment in different ways to advocate their policy preferences. The final outcome depends on factors such as policy supporters' hierarchical position, their persuasive or canvassing abilities, their expertise in the issue area, and how they invoke external factors to shape perceptions of others. Policy outcomes would also be influenced by the degree of organizational change required for their implementation: the greater are the predicted changes, the stronger are the incentives for the "losers" to oppose policy adoption. Consequently, the likelihood of policy adoption decreases.

In examining beyond-compliance policies, the book first employs the neoclassical theory: can the profitability of a beyond-compliance policy be assessed by employing capital budgeting? If this theory does not hold (that is, capital budgeting was either inapplicable, or, if applicable, it was not employed), then I turn to power-based or leadership-based theories. Policy processes marked by imposition are classified as power based, and the ones marked by induced consensus as leadership based. The key actors, policy supporters and policy skeptics, are identified and their positions in the hierarchy, and their strategies and logics for supporting or opposing a policy are examined. Since preferences are inferred from behaviors, the book does not examine why policy supporters or policy skeptics have certain preferences. However, it seeks to understand whether policy adoption requires significant levels of organizational changes that upset the status quo, thereby creating incentives for "losers" to oppose a policy. It also examines how factors external to firms support or impede the efforts of policy supporters.

In summary, the theoretical contributions of this study are fourfold. First, it highlights the inadequacy of the neoclassical theory in explaining why firms selectively adopt Type 2 policies. Second, at a broad level, it argues that "agents" have some (not complete) autonomy in pursuing beyond-compliance policies; external "structures" alone cannot provide fully specified explanations. Third, it focuses on the important role of power-based and leadership-based processes in shaping the policies of firms. It argues for "bringing back leadership" in the study of political economy. Further, the book integrates insights from sociological institutional theory and stakeholder theory (that focus on pressures external to firms) with leadership-based and power-based theories. Finally, since the conclusions of this book are generalizable to other issue areas where firms adopt Type 2 policies (often subsumed under social policies), it outlines important questions for future research.

Research designs and methods

At an empirical level, I focus on two firms – Baxter International Inc. and Eli Lilly and Company – and study their key environmental programs during 1975 to mid 1996. Both Baxter and Lilly are multinational enterprises (MNEs). Since MNEs are important economic actors, they have critical roles in environmental policymaking and implementation (Walters 1973; Pearson 1985; World Commission on Environment and Development 1987; Leonard 1988; World Bank 1992; Schmidheiny 1992; Choucri 1993; Jaffe, Peterson, Portney, and Stavins 1995; Prakash, Krutilla, and Karamanos 1996). Therefore, one objective of many environmental policies is influencing the environmental performance of MNEs. This requires an understanding of how MNEs make environmental policies. Unfortunately, there is little literature on this subject as most environmental policy scholars treat MNEs (or any firm for that matter) as unitary actors.

This study focuses on environmental policymaking in the US operations of Baxter and Lilly. It does not study environmental policymaking in their subsidiaries outside the US. For most MNEs operating in industrialized countries, compliance with domestic environmental regulations is often a non-issue though previously many have resisted complying with laws. I attribute such compliance to stringent laws specifying significant civil and criminal liabilities, relatively serious implementation of environmental laws by regulatory bodies and the courts, active monitoring by environmental groups and local communities, and pressures from employees to "go green." The battle within most firms is now being fought in a different arena: to what extent, if at all, should firms go beyond minimum regulations?

Why Baxter and Lilly? According to the largest ever survey of MNEs' environmental programs, these policies are significantly influenced by MNEs' line of business, sales volume, and home country (UNCTAD 1993). These factors are briefly discussed below.

The line of business The high-risk industries as well as the "sunrise" industries have the strongest environmental programs. High-risk industries such as oil and chemicals have extensive environmental programs because a single industrial accident can inflict significant costs on them. Since sun-rise industries such as electronics, biotechnology, and specialty chemicals have quick product obsolescence, they replace their capital equipment in short cycles. Consequently, they are afforded opportunities to install state-of-the-art, resource-efficient technologies. Further, their high profitability provides them with resources for investing in environmental programs that often have long gestation lags.

The size of MNEs Large MNEs (sales of $4.9 billion and above) have more comprehensive environmental programs than the smaller MNEs because they can tap economies of scale on such expenditures.

The home country of the MNE The scope and content of environmental practices vary significantly across regions. The UNCTAD survey notes that:

[P]robably the nature of the regulatory environments in the home country of the corporation explains variations. . . . The tendency of Asian corporations [that is, Japanese] to view EH&S [Environmental Health and Safety] activities as business opportunities could be related to the fact that Japanese EH&S policy is formulated to a large extent by the Ministry for International Trade and Industry and not the Environmental Agency. The relatively low utilization of EH&S policies and practices in Europe is probably related to the fact that European environmental regulations tend to rest on administrative enforcement and cooperation between industries. On the other hand, United States' environmental regulation has traditionally been described as adversarial and aggressive, and seems to have encouraged the TNCs [transnational corporations] to establish EH&S procedures to minimize liabilities. (1993: 93)

I have controlled for the three factors identified by the UNCTAD report. Baxter and Lilly share the following characteristics:

(1) Their annual sales exceed $4.9 billion ($9.3 billion for Baxter and $6.7 billion for Lilly in 1995).[3]
(2) They are in the health-care industry.
(3) The United States is their home country.

In addition, these firms are:

(4) significantly globalized (in 1995, the non-US operations accounted for 42.5 percent of Lilly's sales and 28.2 percent of Baxter's sales); and
(5) formally committed to adopting beyond-compliance environmental policies (Baxter 1994a; Eli Lilly 1994a).

At an empirical level, this book examines ten cases of Type 2 policymaking: four common to both firms (underground tanks, 33/50, ISO 14000, and environmental audits), and one each idiosyncratic to them (Responsible Care to Lilly and green products to Baxter).[4] The cases that are briefly described below pertain to policymaking during 1975 to mid 1996.

Underground storage tanks

Underground storage tanks can contaminate soil and ground water creating significant clean-up costs. Consequent to the EPA's regulation in 1985, both Baxter and Lilly removed their existing single-walled underground tanks and installed new tanks that have significant beyond-compliance features. I examine why these firms invested huge amounts of money in beyond-compliance features: about $10 million for Baxter and $30–40 million for Lilly.

Toxic Release Inventory and EPA's 33/50 program

Both Baxter and Lilly took significant beyond-compliance initiatives to reduce their releases of chemicals listed under the Toxic Release Inventory program (TRI). Lilly has invested about $80 million for reducing its releases of TRI chemicals and Baxter has invested about $10

---

[3] On October 1, 1996, Baxter announced that it had reorganized itself into two corporations: Baxter International Inc. and Allegiance Corporation. Baxter International Inc. focuses on developing medical technologies and Allegiance Corporation focuses on supplying medical and laboratory products (Baxter 1997). Since I am studying environmental policymaking in Baxter and Lilly during 1975–mid 1996, Baxter's reorganization does not affect my research design or analysis.

[4] In chapter 5, external and internal environmental audits as well as Phase I and II of Responsible Care are examined separately.

million for reducing its releases of TRI chemicals, air toxics, and chlorofluorocarbons.

33/50 is a voluntary beyond-compliance program launched by the EPA in 1991. Firms are encouraged to commit to reducing aggregate releases of seventeen specific TRI chemicals, 33 percent by 1992 and 50 percent by 1995 with 1988 as the baseline. Both Baxter and Lilly are charter members of this program and both have exceeded their 1995 objectives.

Chemical manufacturers association's responsible care

The US Chemical Manufacturers Association (CMA) launched Responsible Care in 1988. Under this program, the CMA's member firms are asked to adopt a series of beyond-compliance policies. This case focuses on Lilly only. After initial hesitation on some aspects of this program, Lilly adopted Responsible Care and now is a show-case example of its successful implementation.

"Green" products

Both Lilly and Baxter have adopted a variety of beyond-compliance policies to "green" their manufacturing operations and management systems. However, only Baxter markets green products, the ones that explicitly promise environmental protection as one of their benefits. Given the nature of Lilly's business of manufacturing and marketing ethical or prescription drugs, green products have little business rationale.

Environmental audits

Though there is major controversy over granting attorney–client privilege to environmental audits, both Baxter and Lilly have established strong internal audit programs. In addition, Baxter invites external auditors to evaluate its environmental programs. In 1991, Arthur D. Little was invited to help in defining the state-of-the-art environmental standards and in evaluating whether Baxter's environmental program met those standards.

International Organization for Standardization's ISO 14000

The ISO 14000 series specifies beyond-compliance management systems. These standards have been sponsored by the International Organization for Standardization, a Geneva-based non-governmental organization. ISO 14000 could be viewed as an industrial code of practice that needs to be certified by external auditors. This certification is done at facility level. Currently, such certification is estimated to cost about $20,000 per facility. Neither Baxter nor Lilly have mandated that their facilities should have the ISO 14000 certification; they have adopted a wait-and-see policy. This could be attributed to their extant investments in other industrial codes (Responsible Care for Lilly and the state-of-the-art program for Baxter), and meager perceived gains from switching over to ISO 14000.

Information on these cases was gathered from the following sources: interviews with managers (both in-service and retired), attendance as an observer in meetings of various environmental teams, review of published as well as unpublished documents, and professional journals. Most managers in these firms have been extremely cooperative in sharing information and have not attempted to influence my interpretation of events. However, to maintain confidentiality of my sources, this book does not identify them in any manner, except when quoting from a published document.

Case selection

In examining the above cases, I define the dependent variable as the adoption or non-adoption of Type 2 policies, and the independent variables as factors internal and external to firms. The internal factors include: whether a policy required up-front capital expenditure or whether it involved establishing management systems; the level of expenditures; and the degree of organizational change required to implement a policy. Some of these policies involved significant capital expenditures (underground tanks and 33/50, in particular) and could therefore have been subjected to capital budgeting. Other policies involved establishing management systems (Responsible Care and ISO 14000) whose financial impact cannot be quantified. In addition, the degree of organizational change required for implementing such policies also varied: "minimal" for underground tanks and "significant" for external audits.[5]

To understand the roles of external factors, the book focuses on the managerial perceptions (and how they were shaped) of the abilities of such organizations to impose excludable costs or provide excludable

---

[5] Minimal changes involve only operational-choice level while significant changes involve collective- and constitutional-choice levels. I discuss the three levels of institutional analysis (operational-, collective-, and constitutional-choice) in chapter 5.

Table 1.2. *Cases and their dimensions*

	External factors encouraging or discouraging this policy	Time period	Scope
Underground Tanks	Encouraged by Environmental Protection Agency	Mid 1980s–late 1980s	Firm
TRI and 33/50	Encouraged by the Environmental Protection Agency	Late 1980s–early 1990s	Manufacturing industry
Responsible Care	Encouraged by Chemical Manufacturers Association	Late 1980s	Chemical industry
"Green Products"	Encouraged by health-care providers, such as hospitals	Early 1990s	Firm
Environmental audits	Discouraged by the Environmental Protection Agency; encouraged by state environmental agencies	Early 1990s–mid 1990s	Firm
ISO 14000	Encouraged by a Geneva-based non-governmental organization	Mid 1990s	All Industries

benefits to firms and to individual managers. The following external factors are examined: governmental agencies (the EPA), non-governmental international organizations (the International Organization for Standardization); industry-level associations (the CMA), and customers (hospitals).

As shown in table 1.2 above, these cases also pertain to different time periods: from the mid 1980s (underground tanks) to mid 1996 (ISO 14000). They also represent policy initiatives at different scales of aggregation: specific to a firm (underground tanks), specific to chemical industry (Responsible Care), specific to manufacturing firms across industries (33/50), and impacting virtually all firms in the economy (ISO 14000).

Following King, Keohane, and Verba (1994), I have selected the cases to ensure variations on independent variables. They also advise that with a small sample size, researchers should consciously ensure variations on the dependent variable as well. As a result, the book also examines four cases of non-adoption (ISO 14000 in Baxter and Lilly; external audits in Lilly, and Phase I of Responsible Care in Lilly) though it primarily focuses on cases where Type 2 policies were adopted.

Organization of the book

The study is organized into five chapters, including this introductory chapter. Chapter 2 presents the theoretical foundation, focusing on the new institutionalist perspective and power-based and leadership-based theories of firms. New-institutionalists focus on two broad sets of questions. First, how do institutions evolve and, second, how do institutions affect collective outcomes? This book focuses on the first question only: how do Type 2 policies – a specific genre of institutions in the context of firms – evolve and how can power-based and leadership-based policies explain their selective adoption? Chapter 3 provides a brief overview of the activities of Baxter and Lilly, and describes and compares the evolution of their environmental programs from 1975 to mid 1996. Chapter 4 examines ten cases of Type 2 policies and explores processes leading to their adoption or non-adoption. These processes are examined by employing power-based and leadership-based theories. Chapter 5 discusses the theoretical and policy implications of this book, its limitations, and issues for future research.

2   Environmental policymaking within firms

This book seeks to understand why firms selectively adopt Type 2 policies. At a broader level, it examines the evolution a specific genre of institutions – firm-level environmental policies. To explore the research question, it departs from neoclassical economic theory and employs a new-institutionalist perspective. Instead of treating firms as unitary actors, it views them as composites consisting of multiple managers. The *ultimate* unit of analysis is the individual manager within the firm.

Managers may have different preferences for Type 2 environmental policies: some – policy supporters – prioritize environmental objectives over a simple effort to maximize quantifiable profits while others – policy skeptics – do not. Once we acknowledge that some managers may pursue environmental objectives that do not or cannot meet the (quantifiable) profit criteria, firms' policies can no longer be explained as their passive responses to external stimuli, in particular, market signals and governmental regulations. Rather, policies *also* reflect intra-firm dynamics and it is difficult to predict *ex ante* whether policy supporters or policy skeptics will always prevail. The final outcome depends on a variety of factors such as policy supporters' hierarchical position, their persuasive or canvassing abilities, their expertise in the issue area, and how they invoke external factors to shape perceptions of policy skeptics. Policy outcomes are also influenced by the degree of organizational change required for the policy implementation: the greater the predicted change, the stronger the incentives are for the "losers" to oppose policy adoption. Consequently, the likelihood of policy adoption decreases.

To understand the intra-firm processes for Type 2 policies, I employ the theoretical tools of new institutionalism and integrate them with insights from sociological institutional theory and stakeholder theory. New-institutionalists emphasize the role of a broad set of institutions in shaping individual incentives, thereby impacting collective action (Bates 1983; Arrow 1987; North 1990; Ostrom 1990; Eggertsson 1990; Furubotn and Richter 1991). In the new-institutionalist tradition, I identify two relevant theories of firms: power-based, and leadership-based.[1] Though these theories view firms as composite actors, they identify different processes of internal policymaking. Individually, both theories illuminate specific dimensions of the research question but neither explains all aspects.

New institutionalism

New institutionalists integrate insights from neoclassical economics, historical institutional economics, political science, and sociology by examining the dynamics between economic and political behaviors in a given institutional context. The central questions addressed by new-institutionalists are twofold: first, how do institutions evolve in response to individuals preferences, endowments, and strategies; and second, how do they impact collective-level outcomes? *The focus of this book is on the former: why and how specific kind of firm-level institutions – Type 2 policies – evolve.*

New institutionalists distinguish institutions from organizations. Institutions are enforced rules, formal and informal, about what actions are required, prohibited, or permitted (Ostrom 1986) while organizations are collections of physical actors (North 1990). For example, courts of law are organizations, whereas laws themselves are institutions. Similarly, a firm is an organization while its policies and the rules governing the actions of its managers are institutions. Managers (unitary actors) of firms (composite actors) operate within constraints set by the institutions that are both internal and external to firms. Since institutions are the mediating variables between individuals and collectivities, new-institutionalists analyze various realms of collective action (markets, firms or commercial bureaucracies, non-profit bureaucracies, and governmental bureaucracies) using the same theoretical perspective.

The fact that institutions are influential does not necessarily imply that actors passively respond to "structures." Unlike Hayek (1955) who treats institutions (specifically, markets) as spontaneous outcomes, new institutionalists view institutions (structures) as purposive artifacts. Even though institutions matter, agents can potentially create or modify some structures at one point in time which, in turn, constrain their behavior at another point in time. For this reason, Snidal (1995) refers to institutions as endogenous parameters. As a new institutionalist, I focus on induced

---

[1] As explained in the introductory chapter, the third theory – transaction cost – is not considered because it is not relevant to the research question.

and imposed cooperation regarding Type 2 policies and impediments to it.

In undertaking institutional analysis, new institutionalists employ three assumptions: methodological individualism, pursuit of self-interest, and bounded rationality. These are discussed below.

Methodological individualism

Individuals are viewed as the *ultimate* units of analysis. Buchanan (1962) emphasizes the distinction between individualism as a method of analysis and individualism as a norm for organizing society. The former suggests a focus on the individual as a decision maker: what choices he/she faces and what opportunities he/she has to solve its problems.[2] Consequently, methodological individualists do not reify collective entities such as firms or governments; they treat them as composite actors. Jensen notes that:

Organizations do not have preferences, and do not choose in the conscious and rational sense that we attribute to people. Anyone who has served on committees understands this fact. Usually no single person on a committee has the power to choose the outcome, and the choices that result from committee processes seldom resemble anything like the reasoned choice of a single individual. (1983: 327)

Methodological individualism is a key pillar in my attempt to understand policymaking in firms. I trace the evolution of Type 2 policies (institutions) of firms (organizations) to preferences, endowments, and strategies of managers. As a result, I view Type 2 policies as purposive artifacts, and institutional change (policy adoption or non-adoption) as a process that is affected by the extant structures as well as the preferences, endowments, and strategies of managers (Schelling 1978; Arrow 1987; North 1990; Ostrom 1996). Actors, however, cannot always modify or create new institutions as institutional change is affected by a variety of factors such as path dependency, transaction costs, and the rules used to change rules. Thus, and not surprisingly so, policy supporters fail to have their firms adopt Type 2 policies.

Both the neoclassicalists and the new institutionalists assume methodological individualism. However, the former contend that markets evolve

---

[2] Buchanan notes: '[I]ndividualism *as a method of analysis* and individualism *as a norm for organizing society* . . . [the former] suggests simply that all theorizing, all analysis, is resolved finally into considerations faced by the individual person as a decision-maker . . . analysis reduces to an examination of his choice problem and of his means or opportunities for solving this problem. To this approach is opposed that which starts from the presumption that some unit larger than the single person . . . is the entity whose choice problems are to be examined' (1962: 315).

spontaneously. The Austrian School, particularly Hayek (1945), also rejects "constructive rationalism"; he suggests that institutions cannot be constructed based on a rational design. Evolutionary economists also do not treat institutions as purposive artifacts (Nelson and Winter 1982). Similarly, sociologists or the interpretivist institutionalists contend that institutions evolve exogenously to the actors (Keohane 1988; Ostrom 1991). Thus, new institutionalists view some institutions as potentially purposive artifacts while interpretivists do not.

Pursuit of self-interest

Actors, alone or in groups, are assumed to pursue their self-interests subject to institutional constraints.[3] Both the policy supporters and the policy skeptics are self-interested actors serving their respective material and non-material (ideological or moral) objectives. Williamson (1964) argues in one of his earlier works that managers maximize utility functions that include variables such as status, salary, and prestige. It is conceivable that environmental managers have a vested interest in having Type 2 policies adopted since this often leads to increased budgets for their department and increased headcounts. This, in turn, creates promotional opportunities for them and also increases their prestige within the organization.

However, actors may not always succeed in gaining the highest level of joint net benefits and institutions may therefore generate pareto-inefficient equilibria, the Prisoners' Dilemma game being an eloquent example. Institutions may also serve the interests of dominant actors and represent distributional conflicts (Libecap 1989; North 1990; Knight 1992).

Bounded rationality

Actors are assumed to be boundedly rational: "intendedly rational but limitedly so" (Simon 1957: ix). Bounded rationality reflects information scarcity at the level of an individual actor. In the study of social systems, Eggertsson (1996) identifies three types of information issues: (1) prob-

---

[3] Principal–agency notions are critical in understanding behaviors of composite actors (Berle and Means 1932; Manne 1965; Ross 1974; Jensen and Meckling 1976). This book does not examine principal–agency issues relating to Type 2 policies. Also, some scholars do not view principal–agent conflicts to significantly impact working of firms. Manne (1965) argues that markets for managers substantially control agency costs. Demsetz and Lehn (1985) cite evidence suggesting that organizations with widely dispersed ownership (high principal–agent conflicts) as well as with concentrated ownership (low principal–agent conflicts) have comparable levels of profits.

lems in creating and gathering information; (2) neurological constraints in absorbing and processing information; and (3) imperfect models of the external world that limit useful applications of processed information. The first is a systemic constraint since it is practically impossible to collect perfect and complete information on every subject. In contrast to issue (1), issues (2) and (3) signify individual-level constraints; therefore, I am of the view that only (2) and (3) together reflect bounded rationality.

For Simon, bounded rationality implies that actors use rules-of-thumb or standard operating procedures (such as capital budgeting) to make decisions in repetitive situations, and *satisfice* rather than *maximize*. Although satisficing may correctly describe human behavior, it is difficult to operationalize. Some new institutionalists therefore interpret bounded rationality as utility maximization under transaction cost constraints (Eggertsson 1990).

Since boundedly rational actors may make decisions under uncertainty, decision-making may be influenced by managerial interactions. When provided with new evidence or alternate interpretations for existing evidence by policy supporters, skeptical managers may revise their assessment of policies. For example, opposing managers may change their assessment of a policy (they may begin to view it as Type 1 instead of Type 2) if a credible actor were to present the existing cost or revenue data.

Nature of goods and services

Adam Smith (1970) viewed the "invisible hand" as a mechanism to harmonize individual and collective interests when exchanges of private goods and services are organized in competitive markets. However, markets may fail when substantial externalities are present. In dealing with such market failures, individual and group rationality may conflict, leading to collective action dilemmas or social traps (Olson 1965; Hardin 1968; Platt 1973; Cross and Guyer 1980; Ostrom 1990). If actors pursue their short-term self-interests, the group arrives at a socially non-optimal situation. A socially optimal outcome can be achieved only if actors suppress their short-term self-interests.

New-institutionalists trace collective action dilemmas to the physical and institutional nature of goods and services (henceforth referred to as products). Products may initially be classified along two attributes: rivalry or subtractability and excludability (Ostrom and Ostrom 1977). Rivalry implies that if A consumes a particular unit of a product, B cannot. Some products may exhibit congestibility: non-rivalry if the number of consumers is below a given threshold, and rivalry if it is beyond that threshold. Excludability implies that it is both technologically feasible and economical for A to exclude B from appropriating benefits once a product has

Table 2.1. *The nature of goods and services*

Excludability Rivalry/subtractability	Easy	Difficult
Rival	(1) Private goods	(4) Common-pool resources
Partially rival or congestible	(2) Club goods	
	(3) Toll goods	
Non-rival		(5) Pure public goods

Source: Adapted from Ostrom and Ostrom (1977: 12).

been produced. Excludability, reflecting the ease of defining and enforcing property rights, is influenced by both physical and institutional factors. As illustrated in table 2.1, based on physical and institutional contexts, products can initially be classified into stylized categories.

Pure public goods are non-rival and it is difficult to exclude actors from appropriating their benefits. In contrast, private goods are rival and easily excludable. Common-pool resources are rival, but it is difficult to exclude actors from appropriating their benefits. Club and Toll goods are partially rival (congestible) and easily excludable.

Public goods are specific to the jurisdictions that provision them. We can think in terms of local public goods (street lights), national public goods (national defense), and global public goods (protecting stratospheric ozone). However, public goods may create externalities for jurisdictions that are not provisioning them.

Categorizing products is important since these categories represent different types of collective action dilemmas. If actors cannot be excluded from partaking in the benefits of a product, they have few incentives to contribute to the product's provision. Difficulties in exclusion create incentives for "free riding" (Olson 1965) leading to under-provisioning of the product. Since markets elicit contributions towards provision through the threat of exclusion, they are ineffective in providing non-excludable products, and some alternative institutional mechanism is required (Ostrom 1990).

Rivalry in consumption leads to a different category of problems; it creates incentives for overuse and rent dissipation (Ostrom, Gardner, and Walker 1994). If a product is highly rivalrous and easily excludable (a private good), then its over-consumption is checked through a rise in its price. However, if exclusion from a highly rivalrous product (a common-pool resource) is difficult or costly, markets do not function effectively, and scarcity does not translate into a higher price. Unless there are counter-acting institutions, there is over-consumption leading to degradation of the resource. Again, non-market institutions are necessary to ensure sustainable use of such resources.

Scholars often assume that governmental provision is necessary for non-rival/congestible goods. However, congestible and easily excludable goods (impure public goods) can be provided by non-governmental organizations as well (Cornes and Sandler 1996). Such goods are of two kinds: toll and club.[4] The former can be provisioned by levying a user toll. Consumers, paying for every additional unit, reveal their preferences, and one can conceivably think of a continuous demand schedule for each consumer.

In contrast to toll goods, we cannot price the discrete consumption units of club goods. As a result, their provision is financed by membership fees. Industry-level initiatives such as the Chemical Manufacturers Association's (CMA's) Responsible Care (chapter 4) and ISO 14000 (chapter 4) are examples of club goods because one cannot price the discrete units of "goodwill" (Commons 1968: 199) benefits for the firms that are generated by such initiatives. Firms will have incentives to pay their membership fees only if such benefits are made excludable. Membership fees can take various forms. Consider the case of ISO 14000. Here the membership fee includes the costs of implementing environmental management systems and hiring experts to certify them. ISO 14000 certification is the "fee receipt" conferring the goodwill benefits. For Responsible Care, the membership fee takes the form of implementing new organizational policies such as community outreach programs. Since the CMA makes it mandatory for members to subscribe to Responsible Care, membership to the CMA constitutes the fee receipt.

Type 2 policies such as Responsible Care and ISO 14000 create several joint benefits. First, by reducing pollution, they create benefits for citizens at large. Since these benefits are akin to public goods, firms cannot charge for their provision. Second, these policies generate goodwill towards firms among regulators, citizen groups, local communities, and financial institutions. In addition to difficulties in monetizing them, these benefits have the characteristics of public goods. Consequently, firms have few incentives to provide them.

"Green products" (chapter 4) may also represent firms' attempts to transform benefits of environmental protection from public to private goods. If consumers are willing to pay a premium for green products (and have the satisfaction of being environmentally responsible), then firms can offset additional costs of adopting Type 2 policies. However, firms do not adopt this route *en masse* since it transfers collective action dilemmas from firms to consumers. Rational consumers may want to enjoy benefits

---

[4] The concept of an "impure public good" was popularized by Buchanan (1965) though its application can be found in works of Tiebout (1956), Wiseman (1957), Pigou (1960) and Olson (1965). I am indebted to Brenda Bushouse for highlighting the distinction between club and toll goods.

of a clean environment (from which they cannot be excluded) without paying for them. If such defections are widespread, then markets for premium green products would remain small, and firms pursuing profits would have few incentives to adopt Type 2 policies.

Type 2 policies also generate goodwill for firms. To transform non-excludable goodwill benefits to club goods, industry-level institutions may be required. Initiatives such as the Environmental Protection Agency's (EPA) 33/50 program, and the CMA's Responsible Care (all chapter 4) fit into this category. Through Responsible Care, the CMA seeks to generate goodwill for the chemical industry, particularly for its members. To ensure that its members do not free-ride, the CMA *requires* that all its members subscribe to Responsible Care. Similarly, firms adopting ISO 14000 management systems will get the ISO certification conferring goodwill benefits to them. This again transforms goodwill benefits from a public good to a club good.

Goodwill benefits could also be transformed to private goods. For example, financial organizations (such as mutual funds, insurance companies, and pension funds) may reward "green firms" that adopt Type 2 policies. However, as Schmidheiny and Zorraquin (1996) point out, since the primary objective of financial institutions is to maximize shareholders' wealth, most of them would probably penalize firms that adopt Type 2 policies. Collective action dilemmas are now at the investor level: rational investors may desire the benefits of a clean environment (from which they cannot be excluded) without paying for them. Since "green investors" constitute only a niche market for investment funds investing in "green securities" only, "green funds" remain peripheral players in financial markets.

Notions of efficiency and capital-budgeting

The concept of efficiency is intrinsic to neoclassical economic theory. Markets are viewed to lead to efficient macro-level outcomes while profit-maximizing firms are viewed to work towards maximizing the efficiency of their operations. The term efficiency has many interpretations and I discuss two relevant notions below. My argument is that no matter which definition of efficiency we adopt, neoclassical theory cannot explain why firms selectively adopt Type 2 policies (Prakash 1999b).

Substantive and procedural efficiency

Since neoclassical theory views firms as profit maximizers, the issue arises as to how do firms maximize profits in their day-to-day activities. Drawing on Simon's (1957) notion of substantive and procedural rationality, this book distinguishes between two notions of efficiency: substantive and procedural.[5] The former suggests *what* policies ought to be adopted while the latter describes *how* they should be adopted.

Theories employing the notion of substantive efficiency assume that managers are fully rational, having full information about the future costs and benefits of a policy. They, therefore, focus on policy outcomes rather than on the processes leading up to them. Substantively efficient firms adopt only those policies that *ex ante* maximize quantifiable profits. What if, as in the case of Type 2 policies, managers do not or cannot quantify profits? Or, if some policies are viewed to serve non-profit objectives? Clearly, any theory based on substantive efficiency offers little help in this regard. It could perhaps be argued that substantive efficiency is an instrumental criterion; this is how firms ought to behave. If treated as a descriptive criterion, we should focus on systemic, not firm-level, outcomes. In the long run, at a systemic level, only those firms that employ the notion of substantive efficiency will survive (Alchian 1950).[6] This may well be correct and I therefore do not dispute the usefulness of substantive efficiency for certain research questions. However, this notion is not helpful in answering my research question that focuses on firms (not the system) and examines the processes leading to adoption or non-adoption of a given Type 2 policy.

I now turn to the notion of procedural efficiency. Instead of focusing exclusively on the final outcome, procedurally efficient managers employ procedures that seek to maximize quantifiable profits. This assumes that such procedures, on average, have tended to produce substantively efficient outcomes. Simply having a procedure in place to maximize profits constitutes "intended efficiency" only; evidence of their success, on average, makes them procedurally efficient. On this count, substantive and procedural efficiency are two facets of the same managerial objective: maximization of quantifiable profits.

Capital budgeting

Since this study focuses on intra-firm dynamics and procedures regarding Type 2 policymaking, the notion of procedural efficiency is the more appropriate to examine whether policy outcomes satisfy the profit maximization criteria. I operationalize procedural efficiency in terms of standard and well-accepted methods of project appraisal, often subsumed under

---

[5] Simon suggested the notion of procedural rationality as a critique of neoclassical theory's notion of substantive rationality.

[6] Alchian's contention is perhaps not falsifiable in that it does not stipulate any *ex ante* criteria.

capital budgeting.[7] To cope with uncertainty about future costs and revenues, and to impart objectivity to investment analysis, managers adopt procedures such as capital budgeting. Predicting profits is difficult due to the bounded rationality of managers as well as uncertainty about variables exogenous to firms. Baumol describes capital budgeting as follows:

Capital budgeting refers to investment decision-making procedures of business firms and other enterprises. The subject encompasses such topics as the selection of projects (which new factories, if any, should the company build), the timing of investment, the determination of the amount to be invested within any given time period, and the arrangements of the financial means necessary for the completion of the project. The calculations which are appropriate for these decisions for the most part derive directly from the theory of capital. . . . However, there is an extremely important limitation which must be emphasized from the very beginning. Imperfect foresight into the future, risk, and uncertainty will for the most part be ignored because economists have not devised really effective methods of taking them into account in the analytical methods. Unfortunately, capital-budgeting is the one subject were we can least afford to abstract from limitations in our knowledge of the future, because, by its very nature, the investment decisions can only be justified in terms of their prospective effects. (1982: 597)

As a tool for investment analysis, capital budgeting (or discounted cash flow techniques) is well established in the Anglo-American system of capitalism (Schall, Sundem, and Geijsbeek 1978; Wicks 1980; Oblak and Helm 1980). The two popular capital budgeting methods are the Net Present Value (NPV) method and the Internal Rate of Return (IRR) method. In the NPV method, managers make projections of costs and revenues emanating from the project over a ten to twenty year period, and discount these cash flows by firms' cost-of-capital. Broadly, cash flows equal net profits plus depreciation. Depreciation is added back to the equation since it is only an accounting cost, and not a cash outflow. Cost-of-capital is the cost of raising funds in the capital market; alternatively, the opportunity cost of investing funds in the next best alternative. If the net present value of the discounted cash flows is positive, such projects are deemed as being potentially profitable, and, as a result, firms consider investing in them.

In employing the IRR method, managers identify the discount rate that equates discounted benefits to discounted costs. If this discount rate is greater than the firms' cost-of-capital (also called the hurdle rate), firms consider projects investment worthy. The NPV and the IRR methods are similar but not identical. Specifically, in the IRR method there can be multiple discount rates equating discounted benefits to costs.

---

[7] In public finance literature, capital budgeting refers to an accounting procedure and not an investment appraisal procedure. In this book, capital budgeting refers to the latter only.

Capital budgeting is important because managers put forward their best estimates of benefits and costs emanating from a particular project. Such procedures enable them to cope with uncertainty, and to establish transparency and impartiality in the project appraisal process. Financial markets perceive this as evidence of firms' commitment to financial discipline. With active transnational mergers and acquisitions markets, managers have incentives to closely monitor the reaction of financial markets to firms' investment decisions. Further, in many firms, compensation of top executives is linked to the price of a company's stock (the deferred payment compensation). Hence, managers have incentives (that is, the "stick" of being taken over by another firm, and the "carrot" of enhancing their earnings) to closely attend to the evaluation of financial markets about their firms' performances. Finance and accounting managers, in particular, are major supporters of employing capital-budgeting to assess project viability. Deviations from this practice are rare, especially if projects involve millions of dollars.

Arguably, there are other methods of investment analysis as well such as full-cost accounting and life-cycle analysis that firms could potentially employ. These methods force firms to take into account long-term social (as well as private) costs, thereby internalizing environmental externalities. I have not examined these alternative investment appraisal techniques for several reasons. First, Lilly and Baxter do not routinely employ any of these methods while they do employ capital budgeting. Further, even as a non-routine measure, Lilly and Baxter did not employ them in making decisions on Type 2 policies that this book examines. Second, the methodologies and operationalization of these techniques are not standardized. Hence, they lack legitimacy, especially with the finance and accounting managers. Third, these techniques may not be compatible with the dominant paradigm about the nature of the firm – the main objective of firms is maximizing shareholders' wealth, an objective that capital budgeting focuses on. This study is arguing that this dominant paradigm is inadequate to explain selective adoption of Type 2 policies. Hence, the existence of alternative methods that are not standardized, that lack wide-spread acceptability within and across firms, and that neither Lilly nor Baxter employ, does not undermine my operationalization of procedural efficiency in terms of capital budgeting.

Capital budgeting is indeed a well-established and widely employed tool of investment analysis which reinforces the argument that firms and their managers attempt to be procedurally efficient. As a result, any deviations from these established procedures (such as adopting Type 2 policies) is puzzling and worthy of examination. The book undertakes this task by "unpacking the firm" and linking policy outcomes to organizational dynamics and politics.

Firms as composite actors

Neoclassical economics does not explain an important puzzle: why do firms arise at all? That is, why and how do managers arrive at "make or buy" decisions? Coase's (1937) seminal article, "The Nature of the Firm," emphasized transaction costs as the key factors in the emergence of firms. Coase criticizes the neoclassical theory in the following terms:

> [It assumes] that the consumer is not a human being but a consistent set of preferences. The firm is defined as a cost and a demand curve. The theory is simply a logic of optimal pricing and input combination. Exchanges take place without any specification of its institutional settings. (1988: 2)

In his Nobel Prize lecture, Coase further observed that:

> [A]n economist does not interest himself in the internal arrangements within organizations but only in what happens on the market, the purchase of factors of production and the sale of the goods and services that these factors produce. What happens in between the purchase of the factors of production and the sale of goods that these factors produce is largely ignored. . . . *The firm in mainstream economics has often been described as a "black box." And so it is.* This is most extraordinary since most of the resources in a modern economic system are employed within firms, with how these resources are used depends on administrative decisions and not directly on the operation of the market. Consequently, the efficiency of the economic system depends to a considerable extent on how these organizations conduct their affairs, particularly, of course the modern corporation. Even more surprising, given their [neoclassical economists'] interest in the pricing system, is the neglect of the market or more specifically, the institutional arrangements to a large extent what is produced, what we have is a very incomplete theory. (1993: 228–9; italics mine)

Coase took the first step in "unpacking the firm." Williamson (1985), building on Coase's ideas, provides a more developed theory to understand why transaction costs arise, and how they impact managerial "make or buy" decisions. Williamson emphasizes incomplete contracting since it is impossible to account for every contingency. He assumes that humans are potentially opportunistic; they pursue "self-interest with guile." He highlights the "fundamental transformation" (from competitive markets to a bilateral monopoly) in many repeated interactions due to "asset specificity" – the productivity of an asset is specific to certain persons and tasks. Asset specificity makes asset owners vulnerable to labors' opportunism, and purchasers and suppliers vulnerable to each other.

Controlling opportunism through external enforcement ("legal central-ism") is costly; actors, therefore, create hierarchies to control it internally at lower transaction costs. Thus, firms are hierarchical governance struc-tures that manage interdependence among economic actors.[8]

Coase, Williamson, and other scholars in the transactions cost tradition have highlighted the shortcomings of neoclassical economics and con-tributed key concepts for unpacking the firm.[9] Transaction cost theories focus on macro-institutional issues such as why firms arise. They are, therefore, inappropriate for this study since it focuses on institutional evolution at a micro-level – why firms adopt Type 2 policies selectively. To answer this question, the book employs power-based and leadership-based theories, both within the new-institutionalist tradition, to "unpack the firm" and to examine intra-firm dynamics on Type 2 policies.

Power-based theories

The term "power" has many meanings. This book employs it to describe the ability of manager A to influence outcomes in wake of opposition from manager B. For example, policy supporters are deemed to exercise power if they force adoption of Type 2 policies although policy skeptics oppose it (Weber 1947; Dahl 1957; Pfeffer 1981; Mitchell, Agle, and Wood 1997).

It is instructive to examine the bases of power and why some manag-ers are more powerful than others. Etzioni (1988) identifies three types of power – coercive, material, and symbolic. Coercive power is based in control over instruments of coercion. For example, some managers may control workers by physically intimidating them. Materially powerful managers control instruments of material well being. For example, a supervisor can control compensation and promotions of a subordinate. Symbolic power suggests that managers control normative symbols that

---

[8] Chandler (1962, 1977) also views modern industrial enterprises as efficiency-enhancing governance structures. His notion of transaction costs is restrictive (or more precise) com-pared to that of Williamson's. Chandler focuses only on the reduction in administrative costs of coordinating resource flows in manufacturing when the production technology permitted mass production of standardized products for national as well as international markets. Further, Chandler does not relate transaction costs to opportunism due to asset specificity.

[9] Williamson is criticized for inadequately explaining how managers operationalize asset specificity and opportunism (Dugger 1983; Dow 1987; Roberts and Greenwood 1997). Critics contend that he provides insufficient guidance for measuring transaction costs. As a result, it is difficult to predict *ex ante* whether managers would favor "make" or "buy" for a specific policy. Critics also suggest that as a key behavioral assumption, opportunism is a bad descriptor and predictor of human behavior. It may also turn out to be a self-fulfilling prophecy (Ghoshal and Moran 1996).

bestow prestige. For example, a supervisor may have the ability to decide on the designations of his/her subordinates. Or, the supervisor could control allocation of work responsibilities; the favored ones could work on prestigious and high-visibility projects. For the purpose of this study, material and symbolic bases of power are relevant. Hierarchically superior managers typically have symbolic and material power over subordinates. Type 2 policies are adopted by a power-based route if policy supporters are either hierarchically superior or can capture the top management, thereby imposing it on policy skeptics. Similarly, hierarchically senior policy skeptics can ensure that Type 2 policies are not adopted.

Power-based approaches are well established in the new-institutionalist perspective. New institutionalists point out that institutions may not always arise to internalize efficiency gains; they may represent distributional conflicts (Libecap 1989; North 1990; Bowles and Gintis 1993). Knight observes that:

[T]he ongoing development of social institutions is not best explained as a Pareto-superior response to collective goals or benefits but, rather, as a by-product of conflicts over distributional gains. (1992: 19)

Similarly, scholars question if firms are merely institutions for organizing value-addition processes. Or, as radical scholars suggest, they are instruments to dominate labor (Perrow 1979) and to facilitate accumulation (Marglin 1974). Edwards (1979) acknowledges the coordination function of firms. However, as coordination can be achieved by tradition and peer groups, he views firms more as instruments for bureaucratic control. Putterman (1984) questions the efficiency rationale for the emergence of firms because labor hiring capital has the same efficiency implications as capital hiring labor.

If firms are organizations to serve the interests of dominant actors, policies of firms should reflect domination of these interests. Adopting Type 2 policies through a power-based route is predicted to generate conflict. Policy skeptics will accept such policies not because they buy into their logic; rather, their acceptance represents the victory of hierarchical superiors (policy supporters) over subordinates. Policy skeptics are not predicted to change their preferences (as reflected in their behavior) about the desirability of such policies. The level of opposition is influenced by factors such as their extent of disagreement with the policy and the fear of retribution. As I elaborate in chapters 4 and 5, if a policy is perceived to reduce departmental budgets, headcounts, or managers' organizational clout, the "losers" could be expected to assume the role of policy skeptics. In such instances, power-based processes are perhaps the only route to have such policies adopted. Power-based processes are therefore predicted to generate covert and overt intra-firm conflicts. Opposing managers may capitulate, but still carry on guerilla warfare. If a major project is adopted that does not satisfy capital budgeting and hurts the interests of mangers, I examine evidence of contestation. Policy skeptics may exhibit their disagreement in a variety of ways such as slowing down policy implementation or even openly sniping at policy supporters during meetings; they may greet reports on policy failures with cynical "I told you so" remarks. I observed such behaviors during my one-to-one interviews as well as in business meetings which I attended as an observer.

Leadership-based theories

Leadership is a highly researched issue in organizational theory.[10] Similar to power-based theories, leadership-based theories also suggest that certain managers play key roles in creating or modifying institutions (Barnard 1938; Follett 1940; Boulding 1963; Miller 1992). Unlike the dominant actors in power-based processes who impose their preferences, leaders are consensus inducers. They have both the political savvy and yet, more ennobling and ethical goals (Lipman-Blumen 1996). Importantly, policy consensus may not arise spontaneously (as in the Hayekian notion of spontaneous cooperation); interventions of leaders are required. Thus, leaders have the ability to build a shared vision and to foster systemic and long-term patterns of thinking through dialogue (Selznick 1957; Senge 1994; Weick 1995).

Leadership-based theories suggest that the presence of leaders is essential for firms to arise and function.[11] This perspective of the nature of firms contrasts with Williamson's who views hierarchies as artifacts to economize on costs of labors' opportunism. Williamson's critics argue that managers may also behave opportunistically (hierarchical failure) by unfairly appropriating profits (Kreps 1990). Firms cannot, therefore, be viewed simply as artifacts to mitigate market failures; the role of leadership is important. Miller notes that:

[10] Leadership theories can be classified into the following categories: Trait theories (Reitz 1987; Katz 1974; Yukul 1981); Group and Exchange theories (Greene 1975; Graen, Novak, and Sommerkamp 1982); Contingency theories (Fiedler 1967); Path-Goal theories (Georgopolous, Mahoney, and Jones 1957; House 1971); Charismatic leadership theories (House 1976; Bass 1985); Transformational leadership theories; and Social Learning theories (Davis and Luthans 1980). For a survey of literature on leadership see Chemers and Ayman (1993), Luthans (1995), and Northouse (1997).

[11] The influence of leaders as agents of change is not limited to the private sector. There is a growing literature on policy entrepreneurship in the public sector (Schneider and Teske with Mintrom 1995).

Managers face short term-incentives to choose inefficient incentive regimes for subordinates. Employees, knowing this, have no reason to trust the employers with information that would make it possible for the employers to make inefficient decisions. Hierarchy is thus a setting for a commitment problem. *The problem can be solved, but only through a set of strategies that are essentially political ... [and a] leadership style projecting trustworthiness* and/or constitutional constraints on the political authority of the hierarchical superiors. (1992: 235; italics mine)

Barnard (1938) suggests that managerial functions extend beyond organizing factors of production, devising appropriate incentive structures to minimize shirking, and writing contracts to guard against opportunism of the subordinates. The distinctive managerial function is to motivate the employees to surmount their narrow, short-term self-interest. Lipman-Blumen (1996) identifies three general styles employed by leaders: direct (managers tightly define their goals and achieve leadership by outstanding performance), relational (managers lead by collaborating, contributing, and empowering people to achieve respective individual goals), and instrumental (managers employ personal relationships and organization politics to achieve their goals, while allowing others to shape the pathways to those ends). This book focuses on instrumental leadership style. Intra-firm dynamics in adopting Type 2 policies are leadership based if they reflect a conscious building of consensus by policy supporters. Since these policies do not increase quantifiable profits, or in some instances create "losers" from organizational change, they could initially be opposed by policy skeptics (or indifferently received by policy neutrals). Due to inter-managerial conflict, policies are often either shelved or adopted by top management imposition (a power-based processes). However, there is a third route as well. Policy supporters may repackage such policies to highlight the long-term benefits, remove the objectionable but unimportant aspects, and persuade policy skeptics to revisit their assessments of benefits and costs. Instead of relying on quantifiable profits as the sole criterion, policy supporters may suggest employing multiple criteria to assess the desirability of policies.[12]

---

[12] In terms of Follett (1940), this could be termed as "integration." According to her, issues of conflict can be dealt with by domination (similar to power-based processes), compromise (achieving middle ground without changing preferences), and integration (involves changing for preferences). Conflict could be constructive, if employed to achieve integration. In my discussion of leader-based processes, integration is being achieved since policy skeptics revise their preferences on the desirability of Type 2 policies. Unlike Follett's conception where both parties change their preferences (what she calls "reciprocal adjustment"), I focus on changing preferences of policy skeptics only.

To employ Boulding (1963) terminology, a leadership-based process could be classified as an integrative response: "[W]hich establishes a community between the threatener and the threatened and produces common values and common interests" (p. 428). In the context of this book, it could be interpreted as a policy outcome that establishes a shared understanding between policy supporters and policy skeptics on the desirability of a Type 2 policy.

How are agendas transformed, deadlocks broken, and what is the role of leaders in this context? Ikenberry's (1993) study of the evolution of the Bretton Woods economic system is instructive on this count. According to Ikenberry, the interventions of policy entrepreneurs at crucial junctures in the form of new ideas and new ways of conceptualizing interests breaks deadlocks, thereby providing a basis for building of consensus. He notes that:

Miracles aside, how does one explain the Anglo-American postwar settlement? Can a simple interest-based argument explain the settlement, or do we need to probe more deeply into the manner in which interests were defined, coalitions were forged, and power was legitimated in the aftermath of world war? I argue that we must probe deeply. . . . A set of policy ideas inspired by Keynesianism and embraced by groups of well-placed government specialists and economists was crucial in defining government conception of postwar interests, building a coalition in support of the postwar settlement . . . these experts and their "new thinking" were important in overcoming political stalemate both within and between the two governments. (1993: 58)

It appears that leaders adopt a variety of strategies to influence policy skeptics. As I discuss in chapters 4 and 5, these strategies often involve impressing the long-term benefits of Type 2 policies upon the skeptics. The presence of a coercive external stakeholder (such as the EPA) that can impose excludable costs or bestow excludable benefits encouraging such policies, imparts credibility to their arguments. Leaders also rely significantly on communicating and sharing information, a tactic that has effectively been used by policy supporters in Baxter. Providing new information and framing existing information in new contexts may stimulate policy skeptics to reassess their preferences and the benefits of a Type 2 policy. They may also seek to create an organizational culture of environmentalism through devices such as regular newsletters and conferences.

The personal credibility of leaders as consistent champions of particular issues, along with their expert knowledge about the issue area, is important.[13] If policy supporters do not have sufficient professional expertise, they could utilize outside experts to make presentations to policy skeptics on the desirability of adopting a policy. Thus, leaders could adopt a variety of strategies to influence the skeptics to revise their assessments of Type 2 policies. I do not expect policy skeptics to change their positions overnight; sustained canvassing by policy supporters is often required. During my one-to-one meetings, one manager admitted that some of them were initially very skeptical about "throwing money

---

[13] Also, the role of technical expertise of bureaucrats in influencing public policy outcomes is well established (Lowi 1969; Allison 1971; Dodd and Schott 1979; Katzmann 1980; Khademian 1992)

down the drain" but the "tree huggers do have a point; this benefits us in the long-term."

In both power-based and leadership-based processes, policy supporters are expected to portray Type 2 policies as serving long-run profit and non-profit objectives of their firms. Although policy supporters may make claims about increases in long-term profits, they provide no quantifiable estimates. Profits no longer remain an "objective" concept whose measurement is invariant across actors. This study is not arguing that established procedures of project appraisal are irrelevant. They matter very much and that is why it is difficult for policy supporters to justify why their pet policy should not be subjected to the formal rules of project appraisal. Importantly, such exceptions occur often in evaluating environmental projects, and this study proposes one way of examining the processes that lead to such exceptions.[14]

The next chapter provides an overview of the activities of Baxter and Lilly, and describes and compares the evolution of their environmental programs from 1975 to mid 1996. It lays out the broad institutional contexts in which these firms made policies on environmental issues. A knowledge of these firms' internal and external contexts is critical for understanding the dynamics of power-based and leadership-based processes, both when they led to the adoption as well as the non-adoption of Type 2 policies. Employing power-based and leadership-based theories, chapter 4 then examines ten cases of Type 2 policymaking: four common to Baxter and Lilly (underground tanks, 33/50, ISO 14000, and environmental audits), and one each idiosyncratic to them (Responsible Care to Lilly and green products to Baxter). Finally, in chapter 5, the findings and conclusions of this book are presented.

---

[14] I disagree with the contention that leadership-based processes are akin to power-based processes where power is exercised subtly in terms of shaping opinion. This issue is discussed in chapter 5.

3   Baxter and Lilly: evolution of environmental programs

The previous chapter outlined a theoretical approach to "unpack" firms for studying their internal policy dynamics on environmental issues. This chapter examines the evolution of the environmental function in Baxter and Lilly. It describes and compares their internal environmental policy-making structures. Both Baxter and Lilly have come a long way in developing and institutionalizing their environmental programs. Baxter's and Lilly's experiences are fairly representative of large US firms which confronted a new set of managerial challenges in the 1970s in the health, safety, and environmental spheres. Though many challenges have yet to be tackled, it is fair to say that most firms have undergone a fundamental transformation on how they relate their functioning and policies to the natural environment. This transformation, however, is uneven within and across firms and can be attributed to factors internal and external to firms. The next chapter focuses on the key events in the evolution of Baxter's and Lilly's environmental programs. These cases are examined by employing power-based and leadership-based theories

Baxter: an overview

Baxter International Inc. is a Deerfield, Illinois-based multinational corporation that develops, manufactures, and markets products and services used in hospitals and other health-care settings. On October 1, 1996, a part of Baxter International Inc. was spun off into an independent firm called Allegiance Corporation. As shown in table 3.1 below, the 1995 net sales of undivided Baxter International Inc. (henceforth Baxter) stood at $9.6 billion, and research and development expenditure at $351 million. Since I examine environmental policymaking in the undivided Baxter International Inc. during 1975–March 1996, the splitting of Baxter into two separate firms does not impact my research design

Baxter had five major business portfolios. The first four now constitute the (new) Baxter International Inc., and the fifth has been spun off as Allegiance Corporation (Baxter 1996a):

Table 3.1. *Baxter International Inc.: Financial highlights (in millions of dollars)*

	1995	1994	Change
Net sales	9,619	9,324	3.1%
Income from continuing operations	435	596	(27.0)%
Return on shareholder's equity	n.a.	17.3%	
Research and development expenditure	351	343	2.3%
Research and development expenditure as a percentage of net sales	3.6%	3.7%	

Sources: (1) Baxter (1994a: 1).
        (2) Allegiance (1997).
        (3) Baxter (1997).

(1) Renal Division provides products and services for patients suffering from chronic kidney failure. Baxter pioneered hemodialysis in 1956 and is now investigating animal-to-human kidney transplants.
(2) Biotechnology Group produces plasma derivatives, biotechnology, and blood-handling products and conducts research to treat blood diseases, cancer, and diabetes.
(3) Cardiovascular Group provides products to fight late-stage heart and vascular diseases. It manufactures heart valves, valve repair products, cardiac monitoring systems, catheters, and equipment used in vascular surgery.
(4) I.V. Systems/International Hospital Divisions manufacturer and distribute intravenous solutions and related products.
(5) Hospital Management Business manufactures and distributes surgical and respiratory therapy products and offers services such as inventory management, customized packaging for surgical procedures, and identification of savings opportunities. This division has been spun off as Allegiance Corporation.

Evolution of environmental programs

Baxter's environmental policies have passed through three phases. Baxter had a state-of-the-art environmental program in the late 1970s and the early 1980s. Its environmental programs began to backslide in the mid 1980s. The late 1980s marked the beginning of the third phase when Baxter revitalized its environmental programs. The top-management mandated that Baxter's US-based facilities should establish a state-of-the-art environmental program by 1992 and facilities abroad by 1996. I elaborate on these phases below.

Baxter's environmental programs had an early start. By the mid 1970s, the senior management had recognized the importance of environmental issues, and there were discussions within the company about creating internal environmental institutions and organizations. In 1976, Baxter (then known as Baxter Travenol Corporation) established its environmental department. Since at that time environmental policies were viewed as having predominantly legal and technical dimensions, Ray Murphy and Don Nurnberg were transferred from the Legal and Corporate Facilities Engineering departments respectively to Corporate Environmental Affairs.

For three reasons, 1977 constitutes a watershed in the evolution of Baxter's environmental programs. First, Murphy and Nurnberg organized Baxter's first-ever company-wide environmental conference. This signaled to the whole organization the intent of the top management to emphasize environmental issues and to encourage the active participation of managers from its various divisions and facilities. An outcome of this conference was that environmental coordinators were appointed for all US facilities. This laid the foundation of the facility-level infrastructure for developing and implementing environmental programs.

Second, to institutionalize the momentum generated by this conference at the top management level, a committee of senior executives, the Environmental Review Board (ERB), was established to oversee and guide company-wide environmental programs. G. Marshall Abbey, Senior Vice-President and the General Counsel, was appointed the chairperson of the Board, a position he held until his retirement in 1992. Abbey, a lawyer by training, recognized the growing importance of environmental issues, and that Baxter needed to proactively establish a strong environmental program. Abbey found allies in other members of the ERB: D.G. Madsen and C.F. Kohlmeyer.

Third, the corporate environmental group developed Baxter's first environmental manual and began auditing facilities' compliance with the laws that were identified in the manual. Manuals are useful tools for auditing performance since they explicate the expectations and responsibilities of various positions in the organizational structure. Consequently, developing a manual was an important step in translating the intent behind creating new decision-making organizations (such as the ERB) into concrete policies.

Over the next few years as the responsibilities of the environmental function expanded, communication bottlenecks began to appear. Some managers felt the need to establish a source of information on new developments in corporate environmental policies; annual updates during the environmental conferences were not sufficient. Further, since new laws were being enacted at the Federal and state levels, there was also a need to have updates on how such laws may impact Baxter. As a result, in 1980, the corporate environmental group began publishing Baxter's internal environmental newsletter, *Travenol Environmental Newsletter*, edited by Ray Murphy.

The late 1970s was a good time for Baxter. Its business was booming, profits were robust, and environmental regulations were relatively uncomplicated and easy to meet. In the early 1980s, the health care industry, facing a cyclical downturn, came under severe financial pressure. In a highly controversial move, both within and outside the company, the senior management responded to this exogenous development by merging with American Hospital Supply Corporation (AHSC) in 1985. It was one of the biggest mergers of that time with both of the firms having sales of about $3 billion each. The preoccupation with this stormy merger meant that Baxter's senior managers had less time to devote to 'soft-areas' such as Baxter's environmental programs. For example, in rushing through the merger, they did not correctly appraise AHSC's environmental liabilities. AHSC did not have a strong environmental program; in contrast to eleven managers in Baxter's corporate environmental department, AHSC had only one. However, Baxter's senior management felt that the strategic importance of merging with AHSC far outweighed AHSC's environmental liabilities.

Soon after the merger, as a part of corporate restructuring, the environmental budget of the merged firm actually decreased in absolute terms. Consequently, Baxter's environmental program began suffering from inadequate human resources. This was further accentuated by attrition.

Unfortunately for Baxter, such budget squeezing and attrition coincided with the increasing number, stringency, and complexity of environmental regulations at the federal, state, and local levels. Environmental issues had begun to have a dramatic impact on the costs of doing business, especially on issues such as real-estate transactions, landfills, and transportation of hazardous chemicals. Local communities and environmental groups had also become very aggressive in monitoring the environmental performance of firms. To add to Baxter's woes, unflattering media reports on toxic emissions from its facilities began appearing, even though such emissions were legally permissible. In his address to the company's annual environmental conference, Vernon Loucks, Baxter's Chief Executive Officer (CEO) admitted that:

But for the most part, since the mid-1980s, Baxter's environmental program has slipped. There are several reasons for that. First, there has been an explosion of environmental regulations. . . . Second, our company has gone through some major changes. Early in the 1980s, the pressure to control health-care costs began to build. . . . Our merger with the American Hospital Supply Corporation in 1985 and subsequent restructuring were necessary responses to these developments. . . . But in addressing these vital issues, the company lost sight of the importance of a good environmental program. We began to take our eye off the proactive measures needed to sustain compliance. Scores of new people were taking on environmental duties at our facilities, but we were not giving them the training and time they needed to do their jobs. (Baxter 1990b:1–2)

Prompted by negative media coverage and litigations by the Environmental Protection Agency, Baxter refocused on environmental programs in the late 1980s. Four key individuals – Senior Vice-President, G. Marshall Abbey; Vice-President, Charles F. Kohlmeyer; Head of Environmental Law, Ray Murphy; and Senior Counsel, and subsequently, Ray Murphy's successor, William R. Blackburn – saw within such external pressures the opportunity to push through a pro-environment agenda within Baxter. They found a sympathetic ear in Vernon Loucks. These managers argued that Baxter must proactively deal with environmental issues by having a state-of-the-art environmental program. Abbey and Kohlmeyer marketed this vision to the top management, and Murphy and Blackburn, to the rest of the organization.

In the late 1980s, taking initial steps towards realizing this vision, Baxter adopted three sets of beyond-compliance policies. First, in 1988, the ERB decided to remove all underground storage tanks from its facilities around the world and replace them with expensive tanks having beyond-compliance features (chapter 4). Second, in 1989, the ERB decided that Baxter should aggressively reduce its emissions of air toxins and chlorofluorocarbons (CFCs). With 1988 as the baseline, the ERB set a target for reducing these emissions by 60 percent by 1992, and by 80 percent by 1996 (chapter 4). Third, the ERB decided to assess the environmental liabilities of AHSC facilities; Arthur D. Little (ADL), a leading environmental consulting firm, was hired to develop a screening protocol (chapter 4). Based on this protocol, ADL audited twenty six facilities and reported that environmental programs were indeed in a state of neglect (Baxter 1990a).

In December 1989, Ray Murphy retired as the head of Corporate Environmental Law. As discussed earlier, Murphy contributed significantly towards institutionalizing the environmental function within Baxter. William Blackburn succeeded Ray Murphy as the head of Environmental Law Affairs, and in 1992 he was named as the Vice-

President of the newly created division of Corporate Environmental Affairs.

Soon after assuming his new responsibilities, Blackburn, along with Ron Meissen of Corporate Environmental Engineering (and their respective teams), began preparing a blueprint to revitalize Baxter's environmental programs.[1] On February 16, 1990, Blackburn presented this plan to the ERB where it was unanimously approved. Under this plan, Baxter would establish a state-of-the-art environmental program; ADL was invited to help in this task. Note that ADL had been hired in 1988 to develop a screening protocol. As I discuss in chapter 4 (environmental audits), the adoption of state-of-the-art standards on an accelerated schedule, and the invitation to ADL, was resisted by some facility- and division-level managers. However, a forceful advocacy of the new policy by senior management, especially Loucks and Abbey, left little room for such opposition to continue. As Loucks noted:

Unfortunately, business hasn't done a good job of acting on environmental issues in the past. At least that's what the public thinks. A recent survey by the Roper Organization found that only one-third of those polled felt that business was meeting its obligations to protect the environment. Companies that try to meet this obligation with words alone will not be taken seriously.

It's tempting to ask: Can't we just lie low, wait for the storm to pass, then go about business a usual? No, this won't blow over, and this is no time for our industry to display a bunker mentality. While survey results may shift from time to time, the public's high interest in the environment will remain . . . (Baxter 1992a: 1–2)

Baxter adopted a two-stage strategy. First, with help from ADL, Blackburn's team defined state-of-the-art standards for firms in similar environmental risk categories as Baxter. In 1989, Blackburn had already identified management systems adopted by facilities with strong environmental programs. The first cut of the state-of-the-art standards was based on these data. Once the state-of-the art standards were defined, ADL audited Baxter's corporate and divisional programs against these standards. Since such standards are dynamic, ADL is invited every three years to redefine these standards and recertify Baxter's environmental programs.

ADL reported that by the end of 1990, Baxter's program at the corporate level had progressed two-thirds of the way towards meeting state-of-the-art requirements; programs at the divisional levels met about 40 percent of the state-of-the-art requirements with some divisions as low as

---

[1] I understand from at least two sources that Bill Blackburn was spotted working on this document even on Christmas Eve!

7 percent and others at 80 percent level. Assessments done by the Corporate Environmental Group suggested that environmental programs at the facility level were about 35 to 45 percent of the state-of-the-art requirements (Baxter 1991a).

Following the ADL report and the new policy, Vernon Loucks mandated that all facilities based in the US, Canada, and Puerto Rico should achieve the state-of-the-art standards by 1993, and facilities abroad by 1996. Consequently, the ERB asked Blackburn and his team to prepare a plan for revamping environmental programs. Further, to monitor progress on various environmental initiatives, they also asked Blackburn to prepare a yearly State-of-the-Program Report for presentation to the ERB.

Following ERB's directive, many organizational and institutional changes were implemented. First, a new environmental manual was issued. This manual outlined the responsibilities, procedures, forms, and audit checklists to help environmental managers, whether at facilities, divisions, or at the corporate office, in performing their functions. Second, to give higher visibility and coherence to environmental function, the Environmental Law section was spun off from the Law Department and established as a separate function: Corporate Environmental Affairs. Further, to facilitate closer interaction between Corporate Environmental Engineering (CEE) and Corporate Environmental Affairs (CEA), their offices were relocated in physical proximity. Third, the major responsibilities to design and implement environmental programs was moved from Corporate Groups to divisional environmental managers (DEMs). As a result, new DEMs were recruited; their number increased from six in the beginning of 1990 to fourteen by the end of 1990 (Baxter 1991a).

Packaging reduction initiatives were the next landmark in the evolution of Baxter's environmental program. This was prompted by a challenge put forth by the Coalition of Northeastern Governors (CONEG) to the top 200 users and producers of packaging in the US. In May 1991, on behalf of CONEG, New Jersey Governor Jim Florio, in a letter to Vernon Loucks, asked Baxter to voluntarily set goals to reduce its packaging wastes and periodically report its progress to CONEG's Source Reduction Council. Accepting the CONEG challenge, Baxter committed to a 15 percent reduction in per-unit packaging weight by 1996 with 1990 as the baseline (Baxter 1991b). The ERB asked Corporate Engineering to work with divisions for establishing numerical goals for facilities and divisions. On this count, Baxter went beyond the CONEG challenge in that CONEG's packaging reduction guidelines did not contain any deadlines

or reduction targets (Baxter 1992b). Baxter's Packaging Reduction Task Force set both target dates and numerical objectives on the following:
(1) eliminate foams made of CFCs;
(2) eliminate/reduce use of inks containing heavy metals such as lead, mercury, and cadmium;
(3) modify Corporate Identity Guidelines for the company logo to allow for the use of recycled paper on office stationary and packaging materials;
(4) apply Society of Plastics Industry and American Paper Institute recycling symbols for appropriate packaging;
(5) require increased use of recycled and recyclable products;
(6) promote sale of single package, multi-product medical kits and reusable shipping containers;
(7) minimize using chlorine-bleached papers and paperboard in packaging;
(8) develop programs to encourage its suppliers to adopt similar packaging reduction packages.

Apart from the specific initiatives, Baxter has also established awards for recognizing individual as well as team performance. Three categories of annual awards have been instituted. First, there are awards for environmental managers in facilities and divisions. Facility-level awards are of two kinds: award for the best all-around program and for the best pollution-prevention initiative. Analogously, there are two division-level awards: all-around best program and pollution prevention. Second, the Erwin Awards, instituted in the memory of Dr. Lewis Erwin (who chaired Baxter's Task Force on environmentally sound packaging) are given to teams that have demonstrated excellence in packaging initiatives. Third, the Merit Awards, are presented to non-environmental managers that have significantly contributed to environmental efforts at the facility, division, company, or community levels.

Baxter has also pioneered corporate environmental accounting by calculating the business impact of its environmental initiatives. This is an important step for retaining the support of internal constituents that may perceive environmental programs as reducing profits. As shown in appendix 3.2, the 1995 Environmental Financial Statement suggests that the total costs of environmental programs for 1995 was $25.2 million while savings were $15.2 million. However, recurring yearly savings due to past environmental initiatives were $72.2 million. As a result, total savings to date due to environmental initiatives are $87.4 million. Thus, green accounting seeks to demonstrate the financial viability of environmental projects. Baxter's pioneering role in green accounting has been recognized by a variety of experts including Stephan Schmidheiny, the Chairperson of the World Business Council for Sustainable Development (Schmidheiny and Zorraquin 1996), and a major player in the 1992 Rio Summit.

Is Baxter's commitment to a state-of-the-art program financially viable and therefore sustainable in the long run? Baxter has very efficiently managed its environmental initiatives. Consider Baxter's expenditure on environmental programs as a percentage of sales. Baxter has succeeded in this as well, as it devoted only 0.3 percent of its 1995 sales to environmental programs. This was considerably less than the 2 percent figure reported by the 1991 survey of 220 companies done by of Booz, Allen and Hamilton and the 1995 study of Cooper and Lybrand that focused on firms with sales greater than $5 billion (Baxter 1995a).

Organizational structure

Since 1993, Baxter has steadily consolidated its environmental programs. Its organizational structure from 1992 to 1996 is summarized below:

Level I: Board of Directors; Public Policy Committee of the Board
Level II: Environmental Review Board (ERB); European Environmental Board
Level III: Corporate Environmental Affairs (CEA); Corporate Environmental Engineering (CEE)[2]
Level IV: Division Environmental Managers (DEM)
Level V: Facility Environmental Managers (FEM)

The Public Policy Committee of the Board of Directors is the highest decision-making body on environmental issues. This committee has two functions. First, it annually reviews Baxter's environmental performance. Second, it reviews the environmental plans of Baxter's operating units. Since its members are not employed by Baxter, this committee also provides an external oversight over the environmental programs.

The ERB is the highest internal organization on environmental issues. The ERB oversees all environmental initiatives, company wide as well those pertaining to divisions only. In early 1996, the ERB had eleven members: nine representing senior managers from different functional areas, and two DEMs. To ensure transparency in its decision making, all Baxter's environmental personnel are invited to attend the ERB's meet-

---

[2] In the new Baxter International Inc. (post 1996), CEE has been merged with CEA. Further, the Health and Safety Function has also been merged with the environmental function. As a result, William Blackburn in now Vice-President of Environmental, Health, and Safety (EHS). As I discuss subsequently, in Lilly, EHS constitutes a single corporate group.

ings which are held once every two months; and some of them actually do so.

The ERB is expected to champion environmental programs and provide top management support to them. As Loucks noted in his address to the conference on "Our Environment: A Healthcare Commitment," held in Arlington, Virginia, on March 10, 1992:

> My final and *most important recommendation is to provide strong top-management support* for your environmental program. Do this visibly, repeatedly, and sincerely. Serve as the main salesperson for the program to middle management. . . . Underscore the need for your managers to allocate appropriate resources. Recognize excellence in environmental performance. . . . Without this support, without this leadership, your environmental programs cannot succeed. *To say it starts at the top is really a cliché. It starts there or it doesn't start at all.* (Baxter 1992a: 10–11; italics mine)

Formed in 1992, the European Environmental Board oversees the implementation of environmental policy in Baxter's European Operations. Its status as independent of the ERB suggests that Baxter perceives environmental challenges in Europe as being significantly different from those in the US and other parts of the world. The head of Baxter's European operations is the *ex officio* chairperson of this committee. The head of Baxter's Corporate Environmental Affairs is also an *ex officio* member of this committee. In addition, as of March 1996, this committee had nine senior managers representing various European operations.

Corporate Environmental Affairs (CEA) coordinates environmental activities throughout the company. It also acts as a resource for divisions and facilities. It prepares the annual environmental performance report, organizes environmental audits, and manages legal issues relating to real estate transactions and superfund liabilities. In addition, it manages strategic initiatives such as estimating net savings from various environmental initiatives and evaluation of the ISO 14000 program (chapter 4). Corporate Engineering works closely with CEA on programs having engineering implications. It has managed key programs such as underground tank removal and toxic emission reduction (both chapter 4).

The DEMs are the main actors in initiating and implementing Baxter's environmental programs. Since they manage divisions' environmental budgets, they have significant influence in planning division-level initiatives. They also prepare their division's annual environmental performance report. This report is key to preparing Baxter's annual environmental performance report. DEM's support is often important for undertaking any company-wide environmental initiatives. The DEMs are also responsible for training the FEMs that report to them. The FEMs are responsible for managing all facility-level environmental programs, whether required by law or beyond-compliance. For example, they are expected to initiate and manage community-outreach programs. They are also required to train non-environmental managers on environmental issues.

In 1995, Baxter had 173 managers performing environmental functions. Since some of these managers had non-environmental responsibilities as well, there were 126 full time equivalents (FTE) environmental managers. Of these, ninety-six were at the facilities, twenty were at the divisions, and ten were at the corporate level. This suggests that Baxter's environmental programs are significantly decentralized, as only ten of the 126 FTEs are at the corporate level (Baxter 1995a).

Importantly, there has been a conscious effort to upgrade the quality of the FEMs; two strategies have been employed for this task. First, since training is often identified as a key factor in imparting skills, Baxter has established training norms: the FEMs of large manufacturing facilities are expected to receive a minimum of eighty hours of training per year and the FEMs of smaller facilities are supposed to receive sixty hours of training per year. This target was met in 1995 since, on average, the FEMs received seventy three hours of environmental training. The second strategy is to retain the FEMs, and thereby lowering FEM turnover. Since environmental laws are complex, it takes time for a new FEM to get accustomed to his or her role. As a result, the FEMs are in a position to initiate and manage beyond-compliance programs only after the adjustment period. The DEMs and corporate groups have made significant efforts to attract and retain managers committed to the environmental function. This has often involved significant personal commitment to train and support the new FEMs.

Eli Lilly: an overview

Founded in 1876, Eli Lilly and Company is an Indianapolis-based multinational corporation that develops, manufactures, and markets pharmaceutical products. As shown in table 3.2 below, its 1995 net sales stood at $6.7 billion, and its research and development (R&D) expenditure at $1 billion.

Lilly focuses on five disease categories: central nervous system, endocrine, infectious, cancer, and cardiovascular. Such focus is important since researching and developing new drugs has become extremely expensive. For reference, in 1995, the average cost to discover and develop a new drug was about $350 million and the average length of time to commercialize a drug was in the range of ten to twelve years (Lilly 1992a,1995b). Such cost pressures have also led to a spate of mergers and

Table 3.2. *Eli Lilly and Company: Financial highlights (in millions of dollars)*

	1995	1994	Change
Net sales	6,764	5,711	18%
Income from continuing operations	1,307	1,185	10%
Return on shareholder's equity	42.5%	25.9%	
Research and development expenditure	1,042	839	24%
Research and development expenditure as a percentage of net sales	15.4%	14.7%	

Source: Lilly (1997a).

acquisitions in the pharmaceutical industry across continents, as no single national market can defray the expensive R&D costs. Instead of merging with other drug firms to become a major player in a wide-range of pharmaceuticals, Eli Lilly has chosen to focus its R&D efforts on five disease categories. These categories represent its core-competencies (Prahalad and Hamel 1990) in which it has the critical R&D capabilities. A recent study commissioned by Lilly also suggested that size of firm *per se* is unimportant, once the firm has critical R&D capabilities in a given disease category (Lilly 1994a).

Let me elaborate with an example. Lilly has chosen to focus on diabetes, a disease in the endocrine category. Historically, it has been a leader in diabetes treatment, having developed and marketed the first insulin product in 1923. Many diabetics find that this disease disrupts their normal daily schedules: they are required to check their blood glucose several times a day with cumbersome instruments; they have to take insulin at least thirty minutes before a meal. Lilly has responded to these unmet consumer needs by developing: (a) a user-friendly electronic device for administering accurate insulin dosages; and (b) an insulin analog, Humalog, that can be taken just before a meal. Thus, instead of providing partial treatments in a wide range of disease categories, Lilly has chosen to provide a wide range of treatments in a few disease categories. As a result, if Lilly has a presence in a disease category then it is one of a market leader. Such a culture of doing few things, but doing them well, is also reflected in its environmental programs.

Evolution of environmental programs

The story of the evolution of Lilly's environmental programs is less dramatic than that of Baxter's. Lilly's environmental programs have progressed slowly but steadily with only occasional lapses. Initially, environmental programs were driven by Lilly's business needs, specifically treating waste-water streams from its bulk manufacturing facilities. Lilly has three types of facilities:

(1) Bulk manufacturing facilities, where it manufactures antibiotics in bulk using three types of processes: chemical synthesis, antibiotic fermentation, and bio-synthesis. Chemical synthesis is intensive in solvent use. Bio-synthesis involves using micro-organisms with recombinant DNA. It does not require using solvents or waste-water in significant quantities.

Antibiotic fermentation is intensive in waste-water use since it involves using living micro-organisms as inputs. Colonies of such micro-organisms are developed in large tanks that ferment solutions of corn and soybean meal, lard oil, and plant starches. The manufacturing process is extremely water intensive. Once the manufacturing is over, the fermented material and waste-water need to be disposed. If it is not done in an environmentally sound manner, this can pollute water streams and harm human and aquatic life.

(2) Fill and finish facilities, where the bulk antibiotics are put into capsules or other final product forms.

(3) Research and Development sites and Technology Centers.

Most of Lilly's early environmental initiatives took place in its bulk manufacturing sites at Tippecanoe and Clinton, both in Indiana. Manufacturing processes in these facilities involve both chemical synthesis and antibiotic fermentation.

Lilly has been a pioneer in developing and adopting waste-water treatment technology. As discussed in the case study on underground tanks (chapter 4), as early as 1952, Lilly hired Robert Lowe to install a state-of-the-art waste-water treatment plant in its Tippecanoe facility. In the 1960s, Lilly began supporting a study conducted by Professor James Gammon of DePauw University on aquatic life in the Wabash River. Since the Clinton and Tippecanoe facilities are located on the banks of the Wabash river, Lilly was interested in documenting the impact of effluent discharges from these facilities on the Biological Oxygen Demand (BOD). Specifically, it wanted to investigate whether the investments in waste-water treatment plants have lowered the BOD levels in the Wabash river. BOD is a critical indicator of hospitability of any waterbody for aquatic life: the lower the BOD levels, the higher is the oxygen available to support aquatic life. Gammon's study suggests that since 1975 the BOD levels in the Wabash river have steadily fallen and the Wabash river has become more hospitable for aquatic life.

The late 1970s marked the beginning of Lilly's crop nutrient program. As suggested earlier, the bulk facilities generate fermented wastes in significant quantities. For reference, between 1987 and 1993, they generated between thirty one and forty two million gallons of such waste annually (Lilly 1995a). Since this fermented waste is rich in nitrogen, it can be used as a plant nutrient. Though this is an environmentally safe way of disposing of this industrial waste, its excessive application can result in the leaching of soil and the formation of nitrates in the water table. By providing this fermented waste free of charge to farmers in the vicinity of the Tippecanoe and Clinton facilities, and suggesting ways for their scientific application, Lilly disposed of the waste in an environmentally safe manner and also earned the goodwill of the farmers.

Programs such as establishing waste-water treatment facilities, supporting the Wabash river study, and organizing the crop nutrient programs were handled predominantly at the facility level with little direction from the corporate office. The need to strengthen corporate environmental resources began to be felt towards the mid 1970s. This was primarily due to the complex nature of new laws and regulations, especially the Resource Conservation and Recovery Act (RCRA) of 1976, and the Comprehensive Environmental Response, Compensation, and Liability Act (CERCLA) of 1980. In 1975, to strengthen the environmental function, Bert Gorman was appointed as the head of Corporate Environmental Affairs. Over the next few years, Lilly increased the managerial strength of its environmental organization, both at the corporate and facility levels: from three full-time equivalent (FTE) managers in 1975, to about twenty-five FTEs by mid 1980s. This increase was due to both new hiring from outside of the company and transfers from other divisions within Lilly.

The year 1989 constitutes a landmark in the evolution of Lilly's environmental programs for many reasons. First, in this year, the Corporate Environmental Legal department conducted Lilly's first internal audit. This audit focused on Tippecanoe facility's compliance with RCRA. Since it revealed certain shortcomings, the senior management approved inducting more managers in the environmental organizations, both at the corporate and facility levels.

Second, in 1989, the Environmental Management Committee (EMC; see below) approved an investment of about $80 million to install Regenerative Thermal Oxidizers for reducing emission of methylene chloride from its Tippecanoe and Clinton facilities (chapter 4).

Third, in this year the Food and Drug Administration (FDA) inspected the Indianapolis facility and identified problems in quality control systems. In response, Lilly restructured its Quality Control/Quality Assurance (QC/QA) organization, with Robert Williams as its new head. Williams was a senior manager holding the rank of Vice-President.

Importantly, environmental affairs was added to Williams's portfolio signaling to the whole organization the increased importance of environmental issues. Don Brannon, from Research and Development, was moved in as the new Director of Corporate Environmental Affairs.

Fourth, in 1989, Lilly formally adopted the Chemical Manufactures Association's (CMA) Responsible Care program. Under this program, firms were expected to implement six Codes of Conduct, including community outreach programs. Since there was internal opposition to implementing community outreach programs, Lilly initially implemented only five of the six Codes; community outreach programs began only in 1993 (chapter 4).

Fifth, the underground storage tank removal program was also initiated in 1989. Under this plan, Lilly removed all underground tanks, and replaced them with above-ground tanks that had significant beyond-compliance features. The total cost of this program was about $100 million with the beyond-compliance expenditures amounting to $30–40 million (chapter 4).

Though firms often adopt environmental programs in response to new laws and regulations, they may also play important roles in influencing the evolution of such laws and regulations. On this count, 1990 was an eventful year in that it marked the onset of a visible role for Lilly in influencing proposed laws and regulations. Lilly's legal department identified problems in implementing the 1990 Clear Air Act Amendments. As a result, Lilly sued the Environmental Protection Agency. Though the two parties eventually reached an out-of-court settlement, this well-researched intervention significantly increased Lilly's credibility in the environmental-legal community as a serious player in air and water regulatory issues. In 1993, Indiana's Governor Bayh asked Lilly to nominate one of its managers to serve on the state task force for identifying appropriate mechanisms to fund Indiana's water and hazardous waste programs. In that year, Lilly's legal department also worked closely with the Indiana Department of Environmental Management on Title V Air Program Rules (Lilly 1993). For example, in 1992, Lilly's managers attended 205 off-site meetings, offered written comments on fifty-six proposed legislations/regulations, and gave testimony in six hearings. The comments and testimonies were given on a wide range of issues such as Voluntary Remediation, Agency Funding, Pollution Prevention, RCRA, SARA, and Asbestos Regulations (Lilly 1992b).

Organizational structures such as the CEA are by themselves insufficient to change behaviors of actors within a firm. Often, clear sets of rules – institutions – are required that lay out responsibilities and sanctions for non-performance. Thus, in 1991, to provide a clear communication to internal and external stakeholders on the objectives of Lilly's environmental programs, Eli Lilly issued its *Environmental Policy and Guidelines* (appendix 3.3). In these guidelines, Lilly reaffirmed its commitment to follow or exceed all applicable environmental rules and regulations. Since this document constitutes a public statement of Lilly's intent, it may also be viewed as a set of standards for evaluating Lilly's environmental programs.

Environmental Policy and Guidelines lays out the macro-objectives of Lilly's environmental programs. It is not a tool to monitor progress on the various individual programs. In the highest positivist tradition, Robert Williams believed that Lilly should be able to measure its compliance with *Environmental Policy and Guidelines*. He argued that though environmental audits were useful tools for this task, an annual report on the overall progress of environmental programs was required. Consequently, in 1992, Donald Brannon and Richard Lattimer, both of CEA, put together and presented Lilly's first *Environmental Annual Report*. Its introduction stated:

The purpose of this report is twofold. The first is to highlight the achievements of the environmental program at Eli Lilly and Company. The second is to be forthright regarding the environmental challenges facing Lilly as an inspiration to employees to take the initiative to improve the environmental performance of the corporation. (1992b: 3)

In its earliest version, the *Environmental Performance Report* was a *de facto* consolidated audit report. However, over time, these reports have become more comprehensive and they now serve as a tool for communicating Lilly's progress on its environmental programs, whether or not it is reflected in environmental audits.

In 1991, two initiatives were implemented to create an environmental ethic within the company. First, a Waste Minimization Conference was organized involving engineers and scientists from the various divisions of Lilly. Participants presented project ideas to minimize waste generation. This conference has now become an annual feature. Further, in 1992, the first Global Environmental Managers conference was organized that involved participation of environmental managers from Lilly's subsidiaries all over the world. Don Brannon played a key role in organizing this conference.

Second, in 1991, an interesting program to "green" non-industrial aspects of Eli Lilly was initiated. The objective was to recycle material used in office environments. Four kinds of materials were identified: white paper, beverage cans, laser printer toner cartridges, and telephone books. Goals were also established for each category. For example, compared to the target of recycling 1.5 million beverage cans, about 1.3 million cans were recycled; compared to a target of recycling 3,100 tons of white paper, about 3,700 tons was recycled. Funds generated from this recycling program are donated to local charities (Lilly 1992b). These kinds of initiatives are important because they instill a kind of environmental ethic among employees who generally do not consider themselves a part of Lilly's environmental programs. In addition, such programs also enhance in-company visibility of environmental managers.

During 1990–1993, the number of FTE environmental managers (predominantly environmental lawyers and engineers), both at the facilities and in the corporate groups, increased from 218 to 340 (Lilly 1992b). This was due to two major factors. First, this was almost forced by the ever-increasing number and complexity of environmental laws and regulations at the federal, state, and local levels. Second, there was a significant expansion in Lilly's bulk manufacturing capacity. Since there are many regulatory issues involved in bulk manufacturing, this required significant increases in managerial strength of the environmental function.

The year 1993 marks the peak of the growth phase of the CEA. First, Robert Williams retired. As suggested before, Williams was a key actor in establishing Lilly's environmental organization and institutions. Since, his successors have been of the rank of a Director,[3] some managers interpret this as an indication of diminished clout of its environmental organization. Further, by 1996, as a result of gradual attrition, the FTE headcount of CEA came down to twenty-one from thirty-five in 1991. Such staff reductions reconfirm suspicion about the declining fortunes of the environmental function. However, this also represented the trend towards decentralization: empowering facilities by locating most of the environmental managers at plant sites.

However, the picture is not as bleak as it seems. In 1995, primarily due to the initiative of John Wilkins of CEA, Lilly began publishing an annual Environmental Health and Safety Report to share information on its environmental programs with external stakeholders. In 1996, Lilly's Tippecanoe facility volunteered to become a pilot site for the CMA's Management System Verification (MSV) program. Responsible Care has been criticized for its lack of external verifiability; firms self-certify that their facilities are implementing the various codes of Responsible Care (chapter 4). In response to this criticism, the CMA is conducting MSV on a pilot-basis, and Lilly's Tippecanoe facility is one of such pilots. Under

---

[3] The hierarchy is: Chief Executive Officer, Executive Vice-President, Vice-President, General Manager, Executive Director, and Director.

MSV, managers belonging to CMA member firms will audit the facility's progress on Responsible Care. Importantly, this is the first time that Lilly's environmental programs will be audited by external actors. As discussed in the case study on environmental audits (chapter 4), unlike Baxter, Lilly does not invite external actors for environmental audits. Hence, the first external audit, even though on a trial-basis, signifies an increased confidence within Lilly about the strength of its environmental programs.

Organizational structure

Though Lilly's environmental function is integrated with its health and safety functions, I will describe only the environmental organization. This organizational structure has the following components:

Level I: Public Policy Committee of the Board of Directors
Level II: Operations Committee
Level III: Environmental Management Committee
Level IV: Corporate Groups: Environmental Affairs, Legal, and Engineering; Environmental Coordinators at sites.

Lilly's Board of Directors has five committees: Audit, Compensation and Management Development, Public Policy, Finance, and Directors and Corporate Governance. The Public Policy Committee of the Board of Directors is the highest policymaking body on environmental issues. Since its members are not Lilly's employees, they provide external oversight over Lilly's environmental programs.

The Operations Committee is the highest internal body on environmental issues and reports to the Board of Directors. Its main function is to align Environmental Health and Safety (EHS) strategy with other strategic objectives of the company. All its members are Lilly employees and it is headed by Lilly's Chief Operating Officer.

The Environmental Management Committee (EMC) is the highest internal body that focuses *exclusively* on EHS issues. It monitors progress on the various environmental programs and approves the environmental budgets that have been proposed by the divisions. The EMC reports to the Operations Committee and its members consist of senior managers representing various functions. Corporate Environmental Affairs (CEA) is a corporate group that coordinates activities required to carry out Lilly's *Environmental Policy and Guidelines*. It reports to Vice-President of Bulk Manufacturing, who in turn reports to Vice-President of Manufacturing. Since most of CEA's managers have had prior experience at facilities, they blend technical and operational level experience with an expertise in regulatory issues. As a result, CEA serves as a credible resource that delivers managerial tools to managers at plant sites. CEA manages the agenda of the EMC and prepares Lilly's *Annual Environmental Report*. Its managers represent Lilly in the CMA. Importantly, CEA's managers monitor proceedings of the Indiana General Assembly that pertain to environmental issues, and lobby on behalf of Lilly. For example, John Wilkins of CEA is a registered lobbyist in the Indiana General Assembly. The CEA also advises Lilly's lobbyists in Washington, DC on environmental issues and, with the help of Corporate Environmental Legal department, prepares written comments on the proposed environmental laws. CEA also manages strategic initiatives such as the development and implementation of the Plant Site Environmental Compliance Listing System, environmental risk assessment, and environmental impact analyses (chapter 4).

Another key corporate group, Corporate Environmental Legal (CEL), is a part of Lilly's legal division. CEL represents Lilly in all legal matters such as superfund issues and real-estate transactions. It is the final authority in interpreting environmental regulations even though CEA may provide the initial advice. Along with CEA, CEL organizes and leads environmental audits of Lilly's sites, third-party contractors, and waste-treatment facilities.

Corporate engineering provides technical input to CEA, CEL, and the plant sites. Plant managers have considerable autonomy in deciding their environmental programs. They propose the environmental budgets with corporate groups often having only advisory powers. This decentralization of environmental function manifests in two other ways: (1) plant-site environmental coordinators report to facility managers; they do not even have a dotted-line relationship with corporate groups; and (2) in 1995, of the 334 FTE environmental managers, only thirty three belonged to Corporate Groups: twenty-two in CEA, eight in CEL, and four in CEE (Lilly 1995c).

To streamline its Environmental Health and Safety (EHS) function, Lilly is implementing a Three-Loop Operations Design Model (Lilly 1996a). EHS activities include a wide range of regulatory and performance-based programs requiring varying levels of expertise. To prevent duplicating functions across managers, an expertise-based division of labor is required that best matches the talents of environmental managers with the needs of the organization. The first-loop managers are expected to focus on activities designed to meet the operational requirements for implementing Lilly's *Environmental Policy and Guidelines*. As a result, most of such managers will be stationed at plant sites. The second-loop managers are expected to focus on tactical activities designed to improve existing management systems and technologies. The third-loop managers are expected to focus on strategic tasks. These managers are expected to possess detailed knowledge of regulatory and technological trends. Typically the second- and third-loop managers will belong to corporate groups.

Baxter and Lilly: a comparison of environmental organization

It is instructive to compare Baxter's and Lilly's environmental organizations. Their environmental organizations have many similarities. First, in both firms, committees representing outside directors are the highest decision-making bodies on environmental issues: the Public Affairs Committee for Baxter; the Public Policy Committee for Lilly. Thus, these firms provide for an external oversight over their environmental programs. However, both these firms have a committee of senior managers (EMC for Lilly and ERB for Baxter) that is the *de facto* highest decision-making body on environmental policies. These committees have representation from various functional areas; this again suggests that both Baxter and Lilly acknowledge the cross-functional requirements of environmental programs.

Second, both firms also have a corporate group (the CEA) to coordinate various environmental programs. Further, in both, the corporate engineering group is distinct from other corporate groups, thereby highlighting the relative autonomy of technical personnel from the "policy types."

However, their environmental organizations also differ on three counts. First, in Baxter, the CEA is a distinct functional area headed by a Vice President who reports to the General Counsel/ERB Chair. In contrast, in Lilly, Corporate Environmental Affairs is not a distinct functional area. It is headed by a Director who reports to the Vice President of Bulk Manufacturing. This suggests that the environmental function has more clout in Baxter than in Lilly. Also, within Lilly there is a potential conflict of interest as CEA organizes and conducts (along with Legal) audits of manufacturing facilities. However, some of these facilities also report to the Vice President of Bulk Manufacturing.

Second, in Baxter, the Corporate Environmental Legal (CEL) department is part of CEA. In Lilly, CEL is a part of Lilly's Law Department and not of Corporate Environmental Affairs. Since CEA managers also work on regulatory issues, there is a duplication in activities between CEA and CEL; this may constitute a source of conflict between these corporate groups.

Third, a key feature of Baxter's environmental organization is the important role played by the Divisional Environmental Managers (DEMs). Perhaps, this reflects the relative heterogeneous nature of Baxter's business operations with divisions representing distinct business portfolios. In contrast, since Lilly has a more focused business portfolio, it has little need to divisionalize to the extent that Baxter has, and, as a consequence, there is no intermediary layer of environmental managers between the corporate groups and the plant sites.

To conclude, this chapter briefly described the evolution of environmental programs in Baxter and Lilly. The next chapter (chapter 4) examines ten cases of environmental policymaking. Four of these cases (underground tanks, 33/50, ISO 14000, and Environmental Audits) are common to both firms, and one each is idiosyncratic to them (green products to Baxter and Responsible Care to Lilly). These cases are examined by employing power-based and leadership-based theories of firms that were presented in chapter 2.

Appendix 3.1

Baxter's Environmental Policy

Baxter's Environmental Policy was adopted in 1990. The policy is applicable to Baxter's operations worldwide.

1. **Environmental Review Board.** An Environmental Review Board (ERB) appointed by the Public Policy Committee of the Board of Directors of Baxter is responsible for overseeing implementation of environmental policy. The ERB will review and decide matters of environmental importance and will make an annual report to the Board of Directors.

2. **Legal Compliance.** Baxter will comply with all applicable environmental laws.

3. **Risk Control.** Baxter will not create unacceptable risks to the environment and will minimize risk to the company from previous, existing, and potential environmental conditions.

4. **Waste Minimization.** Baxter will aggressively pursue opportunities to minimize the quantity and degree of hazard of the waste that results from its operations. It will reduce toxic and chlorofluorocarbon air amission 60 percent by 1992 and 80 percent by 1996, from 1988 levels based on equivalent production.

5. **Environmental Leadership.** Baxter will work to become a leader in respect for the environment. It will establish and maintain an environmental program to be considered state-of-art among the Fortune 500 companies. Baxter will accomplish this goal by 1993 in the United States, Puerto Rico, and Canada, and by 1996 worldwide.

6. **Environmental Coordinators and Managers.** The manager of each manufacturing and distribution facility, and other division and group managers where appropriate, will appoint a qualified environmental representative to coordinate and manage the unit's environmental program. However, compliance with this Policy is not just the responsibility of these representatives; it is the responsibility of every employee and particularly every manager.

7. **Training and Audit.** Corporate environmental personnel, divisions, and facilities will provide coordinated, effective environmental training, awareness and audit programs as appropriate.

8. **Unit Manager Responsibility.** The manager of each unit of the company will assure that the following are accomplished by the unit wherever relevant:

   8-1 Determine the facts regarding generation and release of pollutants from its facilities and responsibly manage its affairs to minimize any adverse environmental impact.

8-2 Develop and implement its own environmental management program to comply with this Policy.

8-3 Select, design, build, and operate products, processes, and facilities in order to minimize the generation and discharge of waste and other adverse impacts on the environment.

8-4 Utilize control and recycling technology wherever scientifically and economically feasible to minimize the adverse impact on the environment.

For more details on the implementation of these policies, see Baxter's Environmental Manual.

Vernon R. Loucks, Jr.
Chairman of the Board
and Chief Executive Officer

Arthur F. Staubitz
Senior Vice President, Secretary.
General Counsel and
Chairman of the Environmental
Review Board

Source: (Baxter, 1995a).

Appendix 3.2 Baxter's environmental financial statement

Estimated environmental costs and savings worldwide ($ in millions)

Environmental costs	1996[1]	1995[1]	1994[1]
COSTS OF BASIC PROGRAM			
Corporate Environmental Affairs and Shared Multidivisional Costs	1.4	1.4	1.3
Auditors' and Attorneys' Fees	0.5	0.3	0.4
Corporate Environmental Engineering/ Facilities Engineering	0.6	0.7	0.8
Division/Regional/Facility Environmental Professionals and Programs	6.6	6.8	7.6
Packaging Professionals and Programs for Packaging Reductions	1.0	2.3	1.6
Pollution Controls – Operations and Maintenance	3.6	3.6	3.3
Pollution Controls – Depreciation	1.4	1.7	2.1
Total costs of basic program	15.1	16.8	17.1
REMEDIATION, WASTE, AND OTHER RESPONSE COSTS			
(Proactive environmental action will minimize these costs)			
Attorneys' Fees for Cleanup claims, NOVs	0.1	0.2	0.2
Settlements of Government Claims	0.1	0.0	0.0
Waste Disposal	3.1	2.5	2.4
Environmental Taxes for Packaging	0.8	0.3	0.0
Remediation/Cleanup – On-site	0.3	0.3	1.0
Remediation/Cleanup – Off-site	0.1	0.6	0.0
Total remediation, waste and other response costs	4.5	3.9	3.6
TOTAL ENVIRONMENTAL COSTS	19.6	20.7	20.7

Environmental Savings
(Income, savings and cost avoidance from 1996 initiatives)

Estimated environmental costs and savings worldwide ($ in millions) (cont.)

Environmental costs	1996[1]	1995[1]	1994[1]
Ozone-Depleting Substances Cost Reductions	0.6	0.5	1.9
Hazardous Waste – Disposal Cost Reductions	(0.2)	0.1	0.4
Hazardous Waste – Material Cost Reductions	(0.2)	0.1	0.2
Nonhazardous Waste – Disposal Cost Reductions	(0.9)	1.3	4.6
Nonhazardous Waste – Material Cost Reductions	1.3	(0.5)	4.6
Recycling Income	5.6	5.2	3.0
Energy Conservation – Cost Savings[2]	2.5	1.0	0.5
Packaging Cost Reductions	2.4	5.6	5.7
TOTAL 1996 ENVIRONMENTAL SAVINGS[3]	11.1	13.3	20.9
As a percentage of the costs of the basic program	74%	79%	122%
Summary of savings			
Total 1996 Environmental Savings[3]	11.1	13.3	20.9
Cost avoidance in 1996 from efforts initiated in prior years back to 1989[4]	93.5	81.3	73.7
TOTAL INCOME, SAVINGS AND COST AVOIDANCE IN REPORT YEAR	104.6	94.6	94.6

Notes:

[1] These amounts have been adjusted to reflect the spin-off of Allegiance operations.

[2] Cost savings for 1996 calculated from data. Amount for 1995 is conservatively estimated. Amount for 1994 is only for US energy-conservation projects involving lighting.

[3] Cost avoidance from initiatives completed in prior years is listed as a separate line item in this report.

[4] No carry-forward savings is included for 1995 and 1996 for reductions of those ODSs that are no longer available to Baxter. This year's report used a more accurate method of calculating this line item, taking into account current-year pricing and compounded growth in business activity. This produced a significant increase in these sums over that which would have resulted from the previous method.
Source: Baxter (1997b).

Appendix 3.3 Eli Lilly and Company

Environmental policy

Eli Lilly and Company's mission is to create and deliver superior health care solutions in order to provide customers around the world with optimal clinical and economic outcomes. This mission requires that the company operate all of its facilities worldwide in a manner that protects human health and the environment.

Environmental guidelines

Eli Lilly and Company intends to carry out its environmental policy with a spirit of continuous improvement in the following ways:
The company will design, construct, and operate its facilities in a manner that protects human health and minimizes the impact of its operations on the environment.

The company will encourage and expect each employee to be environmentally responsible.

The company will provide ongoing education and training to Lilly employees so that they will be prepared to deal with day-to-day environmental responsibilities as well as environmental emergencies.

The company will comply with or exceed all applicable laws and regulations. Where existing laws and regulations are not adequate, the company will adopt its own environmental quality standards.

The company will make environmental considerations a priority throughout the process of developing new products.

The company will encourage and promote waste minimization, the sustainable use of natural resources, recycling, energy efficiency, resource conservation, and resource recovery.

The company will communicate its commitment to environmental quality to Lilly employees, shareholders, vendors, customers, and the communities in which it operates.

The company will recognize and respond to the community's questions about its operations.

The company will actively participate with government agencies and other appropriate groups to ensure that the development and implementation of environmental policies, laws, regulations, and practices serve the public interest and are based on sound scientific judgment.

The company will regularly assess and report to management and the Board of Directors on the status of its compliance with this policy and with environmental laws and regulations.

Adapted from: Lily (1996b).

Firms have indeed come a long way in designing and implementing challenging environmental programs. From resisting and challenging environmental laws and regulations, they now are willing to go
beyond-compliance. However, there exist variations within and among
firms on adopting beyond-compliance policies, specifically Type 2 policies, and this book examines such variations in Eli Lilly and Baxter
International. The previous chapters discussed the theoretical framework
and laid out the broad institutional context in which Lilly and Baxter have
developed their environmental programs. I argued that the neoclassical
economic theory cannot adequately explain why firms such as Lilly and
Baxter selectively adopt Type 2 policies. For this we need to "unpack" the
firm and examine intra-firm dynamics. To do so, I suggested employing
power-based and leadership-based theories in the new-institutionalist
tradition and also drawing upon insights from institutional theory and
stakeholder theory.

In assessing the desirability of an environmental project, managers
often have multiple objectives and varying preferences. All objectives
cannot be collapsed into a single dimension – maximizing quantifiable
profits – that standard procedures of project appraisal require. The book
focuses on the most widely accepted procedure by US-based firms –
capital budgeting. Type 2 policies create a political space for discursive
struggles, where intra-firm dynamics shape managerial perceptions on
the long-term benefits and costs of such policies. Profits are no longer an
objective criterion that is invariant across managers. If Type 2 policies are
adopted, it is due to two types of processes – power-based or leadership-
based. In the former, policy skeptics do not revise their opposition to
Type 2 policies and as a result, Type 2 policies can only be implemented
by top management imposition. There is a clear evidence of dissent. In
leadership-based processes, policy supporters manage to persuade policy
skeptics of the long-term benefits and hence are able to induce cooperation. Importantly, Type 2 policies may not be adopted due to the absence

of power-based or leadership-based processes to affect intra-firm dynamics. That is, policy supporters either fail to capture the top-management and have it impose the policy or they fail to induce cooperation by convincing the policy skeptics of the long term benefits, although non-quantifiable, of a given policy. The conditions under which power-based and leadership-based processes are likely to succeed are examined in chapter 5.

This chapter employs power-based and leadership-based theories along with insights from institutional and stakeholder theories to examine Baxter's and Lilly's policy processes in ten cases. The data for these cases were gathered from many sources including interviews, attendance in meetings, and published and unpublished documents. To maintain the confidentiality of my sources, I do not identify my data source in any manner except when quoting from a published document.

There are six sections in this chapter. Section 1 examines policy dynamics on underground tanks and concludes that leadership-based theories best explain policy adoption by these firms. Section 2 focuses on the Toxic Release Inventory and the 33/50 programs that were again adopted by these firms due to leadership-based processes. Section 3 discusses the Chemical Manufacturers Associations' Responsible Care program in the context of Eli Lilly only. Lilly initially resisted this program due to power-based dynamics and subsequently adopted it due to leadership-based processes. Section 4 investigates policies on "green products" in the context of Baxter only and concludes that leadership-based processes best explain their adoption. Section 5 focuses on environmental audits. Lilly undertakes only internal audits while Baxter invites external auditors as well. Lilly's policy adoption is explained by leadership-based dynamics while Baxter's is explained by power-based dynamics. Section 6 examines why neither Lilly nor Baxter adopted ISO 14000 environmental management standards. It suggests that this was due to the absence of either power-based or leadership-based processes and then discusses factors external and internal to these firms that led to non-adoption of ISO 14000.

1   Underground storage tanks

Baxter and Lilly have committed millions of dollars to replacing single-walled underground storage tanks (USTs) located in their facilities without subjecting this policy to any kind of formal investment analysis. In formulating their UST programs, these firms were confronted with the following issues: should they remove USTs at all, if so, how many of them, over what period of time, with what kind of replacement storage tanks, and should this policy should be limited to US-based facilities only or extended all over the world. The variation in expenditure was significant for Lilly ranging from $30–40 million for the least ambitious (though still in compliance with laws) route to more than $100 million for the most ambitious beyond-compliance route. Lilly adopted the most ambitious route by removing all US-based USTs and installing expensive above-the-ground tanks. Baxter installed double-walled underground tanks which were 15–25 percent more expensive than the EPA acceptable versions. Further, even though not required by law, Baxter installed new tanks in its facilities all over the world. By 1993, Baxter had spent over $10 million on its UST program in North America alone. My conclusion is that leadership-based theories best explain Baxter's and Lilly's response to the UST program.

Why did UST removal become a key policy issue in the mid 1980s? Industrial facilities often use USTs to store liquids in the form of raw materials, work-in-progress, final outputs, and industrial wastes. In the mid 1980s, the EPA estimated that about 70–80 percent of the "regu-lated"[1] USTs were constructed of bare steel, most of them were over ten years of age with approximately one-third being twenty years of age or more. Steel tanks often corrode non-uniformly and eventually leak through small holes known as "holidays" (USEPA 1987). In 1988, the EPA estimated that as many as 25 percent of all USTs were leaking (USEPA 1988a, 1988b).

Leaking USTs can pollute soil and groundwater. In fact, USTs are reportedly the main cause of groundwater contamination in many states. This is alarming since about 50 percent of the US population depends on groundwater sources to meet its drinking water needs (USEPA 1988a). Though not all leaking USTs may pose public health hazards, many of them store chemicals containing carcinogenic substances. Inflammable pollutants such as petroleum from leaking USTs can also migrate through soil to nearby structures creating potential fire hazards.

If a UST leaks, clean-up costs can be substantial: in the range of $500,000–600,000 when both soil and groundwater have been contami-nated and about $50,000–60,000 when only soil has been contaminated (Baxter 1995a). The issue gets complicated since, *ex ante*, it is difficult to

[1] The non-regulated USTs included: (1) farm and residential tanks holding 1,100 gallons or less of motor fuel; (2) tanks storing heating oil used on the premises where it is stored; (3) tanks on or above the floor of underground areas such as basements, tunnels, etc.; (4) septic tanks; (5) tanks for collecting waste or storm water; (6) flow-through process tanks; and (7) tanks holding less than 100 gallons (USEPA 1988a).

Table 4.1. *EPA's time plan for implementing underground storage tank (UST) guidelines*

Type of tank and piping	Install leak detection by	Install corrosion protection by	Install spill/overfill prevention by
Tanks and piping installed after December 1988	At the time of installation	At the time of installation	At the time of installation
Existing tanks installed			
Pre 1965 or unknown	December 1989	December 1998	December 1998
1965–1969	December 1990	December 1998	December 1998
1970–1974	December 1991	December 1998	December 1998
1975–1979	December 1992	December 1998	December 1998
1980–Dec. 1988	December 1993	December 1998	December 1998

Source: Adapted from US Environmental Protection Agency, *Musts for USTs* (1988b: 17).

establish whether a UST will leak at all and when it will leak. On the other hand, as the book discusses subsequently, replacing USTs with above-the-ground tanks or double-walled underground storage tanks (DWUSTs) is very expensive (Baxter 1995a).

In accordance with the 1984 Hazardous and Solid Waste Amendments (HSWA) to the Resource Conservation and Recovery Act (RCRA) of 1976, the US Congress mandated the EPA to create a comprehensive national-level UST regulatory program. USTs were defined as those tanks that have 10 percent or more of their volume, including piping, underground (USEPA 1988a). On May 7, 1985, the EPA issued an interim ban on installing new single-walled USTs for regulated products (petroleum products, wastes, and hazardous substances) and required upgrading existing USTs by installing leak detection, corrosion protection, and spill/overfill prevention systems. As shown in table 4.1, UST upgrading had a phase-in period depending on the age of the UST. For example, though leak detection systems were required to be installed for all existing USTs by December 1993, UST owners/operators could install corrosion protection and spill prevention systems by December 1998. Importantly, the new regulations exempted flow-through tanks, the ones whose output feeds into another tank. Such tanks store liquids temporarily since they are a part of the closed-loop production process. Consequently, leaks in such tanks can be easily identified in the next stage of the production loop to which such tanks are dedicated. About 50 percent of Lilly's USTs could be classified as flow-through and would therefore be exempt from the new regulations. However, Lilly chose not to take advantage of this loop-hole and removed all its USTs.

Baxter and Lilly: policy dynamics

Baxter and Lilly used USTs primarily to store chemicals. Baxter is the world's biggest supplier of intravenous fluids whose production processes are water-intensive. Lilly is a major manufacturer of antibiotics whose production processes, especially in the fermentation stage, are organic solvent intensive. Hence both firms required some sort of storage tanks for their normal business operations. These firms responded by requiring their facilities within the US (also Canada and Puerto Rico for Baxter; Lilly did not have USTs in Puerto Rico) to remove all USTs within a time frame that was more aggressive than what the EPA allowed. Baxter also committed itself to removing USTs in its facilities outside North America by 1995 (Baxter 1988). Policy skeptics were concerned that Baxter was investing far too many resources in the UST program in terms of installing expensive new tanks, and expanding this program to facilities outside the US where such tanks were not required. Further, given the huge financial strain of Baxter's then recent merger with American Hospital Supply Corporation (AHSC), the skeptics questioned the wisdom of such an expensive beyond-compliance program. Baxter, however, chose to adopt the more expensive route: install state-of-the-art tanks and extend this policy to its facilities all over the world.

Baxter implemented its UST program in two stages. In the first stage beginning April 1987, Baxter's Corporate Environmental Affairs Division hired Geraghty and Miller, an outside consulting firm, to assess the degree of soil and water contamination caused by existing USTs (Baxter 1985b). Previously, American Hospital Supply Corporation (that merged with Baxter in 1985) had hired Geraghty and Miller to inventorize their USTs. They reported that Baxter had about 180 USTs in North America. As summarized in table 4.2, a preliminary survey of 77 USTs revealed that 60 percent of these had contaminated subsoil or groundwater (Baxter 1991a: 23).

The UST program was launched soon after Baxter's merger with AHSC. As discussed in chapter 3, this merger raised a storm both within and outside Baxter, particularly regarding the top management high handedness. UST policy supporters therefore realized that, although they had set an ambitious agenda for themselves, they must not create an impression of championing yet another top–down program. Policy supporters therefore sought to consciously involve Baxter's employees in various aspects of the program including identification of the number and

Table 4.2. *Baxter's experience: 77 USTs at 32 sites*

	Percentage of sites	Average cost of clean-up/removal (in US$)
No contamination	41	18,500
Soil contamination only	34	61,000
Groundwater and soil contamination	25	262,000

Source: Adapted from *Baxter Environmental Review* (1990a: 23).

status of USTs. In one of its editions, Baxter's *Environmental Newsletter* noted that:

[R]ecently many of you received a questionnaire about the underground storage tanks (USTs) at your facility. This survey is the first step in the corporate UST program. The purpose of this program is to evaluate all of the USTs within the company. . . . The first phase of this program, the survey is providing information on the tank, the facility, the topography, and the groundwater. This information will be used to develop a prioritization scheme for all USTs within the company. Survey data, applicable regulations, and other information will be used to identify tanks that require a more thorough investigation. The final phase of the project will predict the environmental impact from any leaking USTs. . . . Because of high probability of USTs developing a leak, they pose potential liability. To minimize Travenol exposure, consider whether or not your facility really needs a UST or its current number of tanks. (1986: 13)

In the second stage, Baxter's Environmental Review Board (ERB) earmarked a budget, outlined a time schedule, and set up an internal organization headed by Ron Meissen of Corporate Facilities Engineering to remove USTs. Meissen had been with the company for more than fifteen years and enjoyed considerable credibility for his technical as well as inter-personal skills. Corporate Environmental Affairs and Facilities Engineering were required to regularly submit progress reports to the ERB. This signaled to the whole organization the high priority given by the top management to this program and bolstered the credibility of policy supporters on this issue.

High-priority programs may flounder if implemented hastily or if there is insufficient and irregular sharing of information. Since policy supporters (belonging primarily to Corporate Environmental Affairs and Engineering) were sensitive to this aspect, there was a constant flow of information to facility managers on issues such as progress on program implementation, penalties for violating EPA's guidelines, etc. For example, the *Baxter Environmental Newsletter* reported that:

Table 4.3. *Baxter's progress on UST removal program*

	Number of USTs	Number of USTs in US, Canada, Puerto Rico	Number of USTs in Other Locations
1980	240	200	40
1988	175	140	35
1990	60	30	30
1992	25	—	25
1994	6	—	6

Source: Baxter's *State of the Environmental Program* (1995a: 21).

The first enforcement order to correct alleged violations of underground storage tank rules was issued for not cathodically protecting a bare steel tank as required. Even though no leaks or contamination was involved, United Parcel Service was assessed a $9,000 fine and ordered to remove a 10,000 gallon gasoline tank from a truck terminal in Connecticut.(1988: 20)

As summarized in table 4.3, Baxter made steady progress removing all its USTs in North America by 1990 and in rest of world by 1995.

Lilly also adopted a comprehensive program for removing USTs from its US facilities. In 1986, after the prodding of policy supporters, Lilly's Environmental Management Committee (EMC) began focusing on removing USTs on a priority basis. The EMC met virtually every month in 1986 to develop a strategy on the new UST regulations. Since there was scant information within the company on the number of such tanks, William Wager of Environmental Engineering was asked to compile an inventory of USTs. Mr. Wager went about systematically identifying tanks in the US and non-US locations. He found that the USTs were of various ages and capacities. As shown in table 4.4, Lilly had about 300 USTs worldwide in 1986, of which 275 were in the US. Further, 97 percent of them were in two production facilities in Indiana – Clinton and Tippecanoe. As I discuss subsequently, this also created incentives for Clinton and Tippecanoe facility managers to lobby for removing USTs.

To implement the new UST regulations, Lilly's EMC constituted a task-force consisting of James Vangeloff, David Bowman, and James Lowes. This group was asked to design new storage tanks incorporating state-of-the-art technology, and to prepare budgetary estimates for their installation. All these managers had been trained as engineers and had been working for Lilly for at least ten years. Hence they had the technical expertise and organizational knowledge for this task. Importantly, since they had all worked in Lilly's manufacturing facilities, they had first-hand experience of manufacturing and maintenance operations. As a result,

Table 4.4. *USTs in Eli Lilly*

	1986	1991	1994
US facilities	295*	249	22
Clinton, IN	123	87	20
Tippecanoe, IN	155	155	—
Others	7	7	2
Non-US Facilities	*	24*	24*
TOTAL	295*	273	44*

Note:
* This is a partial list as many non-US facilities did not know or did not report these numbers.
Source: Estimated from table 36 of (Lilly 1995c) and page 24 Lilly (1992a).

they had credibility with the plant site managers and received active help from managers such as Roger Lruswick of the Tippecanoe facility.

Lilly's management asked this team to apply the principles of zero-based budgeting in determining whether a particular UST was required at all or whether it was in operation due to historical reasons. In the course of their work, this team found that about 10 percent of the USTs could be eliminated by combining streams, cutting back on back-up tanks, etc. (Vangeloff 1992). This is not surprising as many engineering projects are routinely over-designed; that is, excessive resources are devoted to performing a task. This represents both "organizational slack" (Cyert and March 1963; Liebenstein 1966) as well as path dependency in technological choices.

As indicated earlier, the new EPA regulations did not apply to flow-through USTs. More than 50 percent of USTs at Lilly could be classified as flow-through process tanks and therefore were exempt from EPA regulations. Hence, there was a debate within Lilly whether all USTs should be removed or only the ones which do not fall into the flow-through category.

There were three other major debates as well that had significant bearing on the overall cost of the UST program; first, should the new tanks be placed above the ground (the more expensive alternative) or underground? Note that the EPA regulations permitted double-walled underground tanks. Second, if placed above the ground, should they be horizontal (the more expensive route) or vertical? Third, should above the ground tanks be placed in an enclosed area? Managers at Lilly were thus faced with the following problem: the costs of adopting the most spartan route (and yet remain in compliance with the new regulations) versus the cost of the most ambitious route were significantly different; instead of spending more than $100 million on the most ambitious route, Lilly could spend as little as $30–40 million. The policy skeptics pushed for the least expensive route since the most expensive route required more than twice the amount of expenditure. They also cautioned the policy supporters against over-reacting to the new EPA regulations.

Lilly chose to remove all USTs, including the ones which could be classified as flow-through process tanks. The new tanks were above the ground but below the grade. The former means that tanks are not buried and soil is not placed against their outer walls. Below the grade refers to the elevation of the tanks; that they are below the ground or grade level. The new tanks were placed on platforms, which in turn were placed in a concrete bathtub-like structure.

Two rows of new tanks were placed in a modular fashion. Unlike other good designs such as buried double-walled tanks, or tanks in a vault filled with stones that relied on instrumentation, this design ensured that tanks could be visually inspected (Vangeloff 1992). The tanks were placed horizontally and not vertically. Though this required more space (and more expenditure), it reduced the risk of accidents and sabotage. I understand that one of the lead characters in a popular TV series of those times had a hobby of shooting at tall structures. Mangers were, therefore, very concerned that vertical tanks may tempt some people to actually try out what was being shown on the TV. On this count, horizontal tanks appeared safer.

The storage tanks were placed in enclosures heated to 40 F. This had two major environmental advantages. First, it protected tanks from rain. Tanks often overfill and rainwater can wash away spilled toxic liquids which pollute water bodies. Enclosures would prevent rainwater from washing off spilled toxins. Second, during winters (which are severe in Indiana, the location of 97 percent of the USTs), facility personnel are reluctant to go out in the open to carry out regular tank inspection and maintenance. Placing storage tanks in enclosed heated areas reduced such weather-inspired shirking although it increased costs by about 5–10 percent. Overall, the newly installed modular USTs cost almost twice as much as other EPA acceptable versions.

Analysis

Power-based explanations are not helpful in explaining policy adoption because the UST program did not cause any significant conflicts either in Lilly or in Baxter. This program was adopted consensually and I attribute this induced consensus to active intervention of key managers who succeeded in convincing policy skeptics of the long-term benefits of incorpo-rating Type 2 features in the UST program. Thus, leadership-based explanations are most appropriate to explain Baxter's and Lilly's responses.

Why and how did the leadership mode succeed? Policy supporters focused on the hostile external environment as a key motivation for the firm to go ahead with the UST program. In face of external adversity, the policy skeptics were amenable to arguments of the policy supporters. Thus, policy supporters invoked the coercive character of the external factors to impact internal dynamics and usher in institutional change. Importantly, since policy implementation did not require significant organizational changes, thereby creating "losers," policy skeptics did not have incentives to resolutely oppose the program.

The 1980s witnessed a flurry of new environmental legislation and an increasingly aggressive posture of regulators and citizen groups toward polluting companies. To use the popular expression, the "adversary economy" (Chandler 1980; Marcus 1984; Vogel 1986) became even more adversarial. Media also portrayed leaking USTs as vivid symbols of cor-porate America's indifference to environmental concerns. Markets also began to demand that firms incorporate environmental concerns in their policies. In particular, firms faced difficulties in getting insurance cover-age for damages caused by normal business operations. Since the 1970s, many general liability policies had been excluding coverage for pollution liability unless caused by unanticipated events. This, in part, was moti-vated by court cases where general liability was interpreted as covering gradual pollution and its impact on humans after long periods of time (NAPA 1986). This created a market for insurance coverage against gradual pollution and many insurance firms stepped in to tap this market. In 1985, in anticipation of the imminent Superfund legislation, the market for gradual pollution liability for chemical (non-petroleum) exposures collapsed; insurance premia skyrocketed by almost 200 percent and eleven of the fourteen major insurance suppliers withdrew from the market. If gradual pollution insurance were available, its liability was often limited to a maximum of only $10 million. Thus, buying insurance coverage for chemical USTs became problematic for many firms. UST policy supporters within Baxter and Lilly used this as an example of significant but not easily quantifiable costs of ignoring environmental concerns.

It is also instructive to examine other potential benefits that a UST program bestows on firms and how they impact managerial perceptions of the desirability of spending millions of dollars for incorporating Type 2 features in the program. Adding beyond-compliance features creates three types of benefits. First, for the public at large, they create benefits of cleaner water and non-contaminated soil. These benefits, having characteristics of public goods, are diffused across beneficiaries. Since firms cannot charge beneficiaries for these benefits, such projects do not increase firms' profits. Consequently, profit-maximizing firms have little incentive to supply such public goods.

Second, beyond-compliance features create goodwill for firms with regulators, local communities, citizen groups, etc. Importantly, this goodwill is an excludable private good accruing only to firms undertaking this policy and imparts benefits in many ways: firms get quicker approval for their environmental permits; regulators consult firms on new laws and regulations and incorporate their suggestions; and regulators treat minor environmental violations leniently. Since such benefits to firms cannot be quantified, managers assessing projects solely on the basis of quantifiable profits and costs are not interested in undertaking such projects.

Third, incorporating beyond-compliance features reduces future liabilities of firms in the event of any UST leaks. Lilly's managers argued for placing all new storage tanks in bathtub-like concrete containers so that the tanks could be visually inspected for leaks.

Though in both Baxter and Lilly, some managers were initially hesitant to commit huge funds for USTs that had no quantifiable benefits, policy supporters eventually won them over. In Lilly, I identify five such policy supporters: Neil Pettinga, Earl Herr, Crandle, Bert Gorman, and Daniel Carmichael. These managers had the expertise, networking capabilities, and personal credibility to influence corporate decision-making. Pettinga was the Executive Vice-President of Lilly with a background in Research and Development. Crandle, an engineer, was a Group Vice-President in charge of Lilly's manufacturing operations and reported to Pettinga. Herr, the successor to Crandle, was an Executive Vice-President of Lilly. Gorman, a biochemist by training with a Masters in Management, was the head of the Corporate Environmental Division; Carmichael is the Deputy General Counsel, Company Secretary, and a member of the Environmental Management Committee. Since these individuals represented a variety of skills, their advocacy of the UST program was credible across functional areas.

In Baxter, I identify four leaders who championed the UST program: G. Marshall Abbey, C.F. Kohlmeyer, Ray Murphy, and William Blackburn, all having high credibility and persuasive skills. Abbey was a Senior Vice-President, the General Counsel, and the head of the Environmental Review Board. He was also a member of Baxter's Management Committee, the highest corporate decision-making body. Kohlmeyer was the Vice-President of facilities engineering, the agency responsible for removing existing tanks and installing new ones. Murphy was the head of Corporate Environmental Affairs until the end of 1989, and the Assistant General Counsel. Blackburn was Baxter's Senior Counsel until the end of 1989, and subsequently the Vice-President of Environmental Affairs. All had been in the company for over fifteen years.

As discussed in chapter 3, Baxter's and Lilly's environmental policies have been significantly influenced by their history of problems with the EPA. Hence, these leaders emphasized that removing *all* USTs and incorporating Type 2 features had significant symbolic value for both the internal and the external constituents of these firms. They invoked their firms' previous experiences with the EPA and State Environmental Agencies to argue for proactive UST policies. Further, they argued that it was in the long-term business interest of the firm to remove USTs. Baxter's *State of the Environment* Report therefore observed that:

Contamination by leaking underground storage tanks used to store petroleum or other hazardous substances poses an environmental risk and liability exposure that is likely to escalate with time. Given the high costs of dealing with contamination from underground tank leaks, the money being spent for tank removals now before conditions worsen is clearly money well spent. (1995a: 21)

Baxter treats every facility as a profit center; revenues and costs are separately calculated for every facility and capital expenditures are financed from facility budgets. As a result, facility managers have incentives to oppose expenditures which do not reduce their quantifiable costs. I therefore expect that these managers would have opposed the Type 2 features of the UST program. Policy supporters in Baxter anticipated this opposition and sought to overcome it by making UST removal a corporate-level program. Hence, expenditures for removing UST appeared non-rival to projects which a particular facility manager was promoting. For a facility manager, this program created concentrated benefits (even though non-quantifiable) and diffused costs. Hence it was a natural candidate for their support.

Further, policy supporters realized that since the favorable stance of the former policy skeptics may not last for ever, it was necessary to have an accelerated program for removing USTs. This was an astute political move as coalitions that push for a given policy are often temporary. *Baxter Environmental Review* therefore reminded facility environmental managers that:

Financial reserve accounts have been established at the corporate level to fund the removal of all tanks. This fund will pay expenses for removal of the UST(s), environmental assessment of the subsurface conditions surrounding the tank(s), and clean-up of any contamination costs that may exist . . . *we do not expect the financial reserve accounts will be available to fund UST removal after 1990, so it is important to finalize UST removal and replacement now.* (1989: 21; italics mine)

In Eli Lilly, the policy supporters (belonging to Legal Affairs and Environmental Affairs) struck an alliance with facility managers. As discussed before, nearly 97 percent of Lilly's USTs were concentrated in its two biggest facilities – Clinton, Indiana and Tippecanoe, Indiana – and facility managers had incentives to replace all USTs with sophisticated above-the-ground tanks since it reduced risks associated with manufacturing operations. Thus, this coalition championed removing all existing USTs (both flow-through process as well as non-flow-through process) and replacing them with expensive modular tanks.

Lilly's leaders also emphasized that historically, Lilly has been on the cutting edge of the UST technology. As early as 1953, Lilly hired Robert Howe to construct a waste-treatment plant in its production facilities. Howe constructed a state-of-the-art treatment plant so that waste resulting from fermentation processes (required for producing an antibiotic – mycelia) could safely be put in landfills. Howe's pioneering efforts created a lot of credibility with regulators who often turned to him for advice. Hence, it was argued that the proposed UST program was in keeping with the tradition of being industry leaders in the UST technology.

Finally and importantly, in both firms, the UST program did not upset the status quo. Although task-specific groups were also set up, policy supporters consciously employed extant organizational structures to implement the new policy. Because it did not threaten their standing within the organization, policy skeptics had few incentives to persistently oppose this program.

To summarize, both Baxter and Lilly invested significant sums in Type 2 features of their UST program. However, such capital expenditures were not subjected to any formal investment appraisal procedure such as capital budgeting. Leadership-based theories best explain the adoption of this policy. Policy supporters did not impose this policy on policy skeptics. They succeeded in convincing policy skeptics of the long-term benefits of adopting the Type 2 features, particularly in light of the hostile external environment. Policy skeptics did not have incentives to steadfastly oppose such policies because they did not perceive themselves as losing from the UST program. Consequent to the persuasions of policy supporters, policy skeptics eventually changed their assessments of the long term benefits and costs, thereby favoring the incorporation of Type 2 features in the UST program.

2   The Toxic Release Inventory and the 33/50 programs

Baxter and Lilly have invested significant sums in Type 2 policies to reduce the releases of Toxic Release Inventory (TRI) chemicals without subjecting such policies to formal investment analysis. Lilly invested in excess of $80 million for installing state-of-the-art equipment for curbing TRI releases even though less-expensive equipment ($40–50 million) meeting compliance requirements was available. Baxter invested more than $10 million for reducing emissions of TRI chemicals and air toxics. Further, both Baxter and Lilly joined the 33/50 program. As indicated previously, only thirteen percent of target firms (and only 64 percent of the Top 600 companies) releasing TRI chemicals joined this program. My conclusion is that leadership-based theories are most appropriate to explain Baxter's and Lilly's response to the TRI and 33/50 programs.

A major challenge for environmental regulators has been to ensure that firms reduce their releases of toxic chemicals and share information with local communities on such releases. To meet this objective, the US Congress enacted the Emergency Planning and Community-Right-to-Know-Act (EPCRA) in 1986 and the EPA proposed the TRI and 33/50 programs to consolidate the achievements of this law. EPCRA was enacted primarily in response to the Bhopal disaster.[2] It requires state and local governments to develop emergency response plans for unanticipated releases of certain toxic chemicals. Section 313 of EPCRA requires firms with manufacturing facilities in the US, as well as certain federal government facilities, to submit annual reports (Form R) to the EPA on the quantities of Section 313 chemicals that they have released into the environment. It also requires the EPA to make this facility-specific emission data available to public. To this end, the EPA has developed a computerized database known as the Toxic Release Inventory (TRI).

Not surprisingly, the initial TRI reports created adverse publicity for firms. Many environmental groups published rankings of leading polluters – "the dirty dozens" – in their states and counties. Managers within firms were also shocked to realize that their facilities were releasing significant volumes of toxic chemicals. Consequently, firms faced pressures from external stakeholders as well as employees for reducing emissions of TRI chemicals.[3]

---

[2] On December 3, 1984, the Bhopal facility of Union Carbide released methyl isocyanide causing deaths of thousands of people living in the vicinity of the facility. This underlines the dangers of living in the vicinity of facilities using, manufacturing, or emitting toxic chemicals.

[3] The EPA has proposed increasing the number of substances that facilities have to report for TRI purposes. The CMA opposes the EPA's proposal. It believes that since the TRI database seeks to provide information to communities on the releases of toxic chemicals into the environment, the proposed additions to the TRI do not fit this profile. Further, the CMA contends that the goal of the TRI program was to facilitate pollution prevention and risk reduction. However, the EPA proposal introduces use reduction though material accounting. This was not a part of the 1987 TRI initiative. The CMA fears that material accounting data may give competitors access to confidential information having significant business implications (CMA 1996b, 1999).

The TRI reporting requirements were the first milestone leading to the 33/50 program. The second milestone was the 1990 PPA (Pollution Prevention Act). Under the PPA, the US Congress directed the EPA to address the lack of attention to source reduction in existing environmental laws. Source reduction can be achieved by modifying equipment and processes, designing new products, or changing raw materials in existing products. To create incentives for firms to reduce the releases of TRI chemicals and to implement the mandate of the PPA, the EPA proposed a program called 33/50 in February 1991. Under 33/50, the EPA encouraged firms with US-based manufacturing facilities to *voluntarily* commit to reducing their releases of seventeen chemicals by 33 percent by 1992 and 50 percent by 1995 with 1988 as the baseline.[4] The rationale for targeting these seventeen chemicals was that: (1) they have significant adverse impacts on human health and the environment; (2) they were used in large quantities by US-based manufacturing facilities; (3) their releases relative to their total usage were high, suggesting inefficiencies in the production processes; and (4) their usage as well as releases could be reduced by employing pollution-prevention technologies and practices (EPA 1994).

The first year of TRI reporting was 1987. Due to these new and extensive reporting requirements in which firms were relatively inexperienced, the 1987 data were not considered very reliable. As a result, the EPA chose 1988 as the baseline year for the 33/50 program (EPA 1994). The 1987 TRI report indicated that more than 16,000 US-based facilities released at least one of the seventeen 33/50 chemicals and a total of 1.48 billion pounds. These facilities were therefore identified as the target group for the 33/50 program. The EPA contacted nearly 10,000 facilities of which 1,300 (13 percent) agreed to participate in this program (Sarokin, 1999). This again indicates that an exclusive focus on external factors under-specifies our explanations of why firms adopt or do not adopt beyond-compliance policies.

The EPA Administrator wrote to the chief executives of firms owning these facilities inviting them to join the 33/50 program. By contacting the chief executives, the EPA sought to encourage a pollution-prevention philosophy in the highest echelons of Corporate America. Firms participating in 33/50 were promised certificates recognizing their participation and

---

[4] The seventeen chemicals identified under the TRI program were: benzene, cadmium and compounds, carbon tetrachloride, trichloromethane (chloroform), chromium and compounds, cyanides and compounds, lead and compounds, mercury and compounds, methyl ethyl ketone, methyl isobutyl ketone, methylene chloride, nickel and compounds, tetrachloroethylene, toluene, 1,1,1-trichloroethane, trichloroethylene, and xylenes (EPA 1994).

Table 4.5. *Major TRI and 33/50 program projects at Lilly*

Year of completion	Project	Cost in millions
1988	Upgrade thermal oxidizer at Tippecanoe	$2
1990	Install plume suppressors for fume incinerators at Tippecanoe	$6
1992	Install regenerative thermal oxidizers at Clinton	$37
1993	Install regenerative thermal oxidizers at Tippecanoe	$40
1993	Methylene chloride emissions reduction projects at Clinton	$1
1994	Install air emission control equipment for bulk pharma manufacturing at Tippecanoe	$9
1994	Connect C9 to the regenerative thermal oxidizer at Clinton	$1

Source: Lilly (1992b: 9).

public recognition through periodic EPA reports and press releases. The EPA promised to measure the success of this program only at the national level; not at a firm level. Thus if a firm did not meet its goals, it would not get adverse publicity. Firms also had the autonomy to focus their 33/50 goals on all/some of their facilities as well as all/some of the seventeen identified chemicals. Emphasis on aggregate releases instead of releases of individual chemicals and facilities provided flexibility to firms for developing 33/50 plans as a part of their overall corporate strategies.

Baxter and Lilly: response to TRI and 33/50 programs

As indicated in tables 4.5 and 4.6, both Baxter and Lilly took significant initiatives for reducing their releases of TRI chemicals and achieving the objectives of the 33/50 program (Currie 1995; Wilkins 1996). Both firms are charter members of the 33/50 program. Lilly has spent about $80 million in its Tippecanoe and Clinton facilities for reducing releases of TRI chemicals, especially the 33/50 chemicals. Baxter has spent about $10 million for reducing releases of TRI chemicals, air toxics, CFCs, and 33/50 chemicals.

Of the seventeen chemicals targeted under the 33/50 program, Lilly used toluene and methylene chloride in significant quantities; the latter accounting for over 70 percent of Lilly's releases (Lilly 1996b). Some scientists claim that methylene chloride is carcinogenic in animals by both oral and inhalation routes, and produces toxicity of the liver, kidneys, and the central nervous system. However, there is no conclusive evidence that

Table 4.6. *Major TRI and waste-minimization projects at Baxter*

Project description	Project location	Cost in million
Install ethylene oxide control devices	Anasco/CVG Mt. Home Cleveland Aibonito Jayuya Singapore Colombia	$4.0
Install ethylene oxide/CFC-12 recovery system	Uden	$0.4
Install capture and control system on part washing equipment	Irvine/CVG Mt. Home Anasco/CVG Anasco/Biotech Castelbar/Swinford	$2.3
Convert to water-based coagulant, and install catalytic converters	Johnson City Kingstree Malaysia Exam Malaysia Surgical	$1.7

Source: Baxter (1995a: 10), table 5.

methylene chloride produces cancers in humans. Nevertheless, Lilly adopted three strategies for reducing the releases of this chemical. First, given methylene chloride's propensity to evaporate quickly, Lilly introduced new manufacturing technologies and practices to improve containment within the facility. Second, to make methylene chloride harmless, it was to be centrally collected and incinerated. Third, Lilly focused on reducing methylene chloride's usage in its new products.

For the first task, Lilly's top management constituted a task force known within the company as the "fix-it" team. This team, led by Paul Wassell, sought to identify all possible leakpoints and valves, a tedious task given that manufacturing facilities have hundreds of miles of piping. At the same time, Richard Lattimer estimated the releases of methylene chloride by using sophisticated material-balancing equipment. This team had an estimate of the aggregate quantities of these leaks and the points where they actually took place. To correct this gigantic problem, the team first experimented with many engineering and system innovations on a pilot basis in the Clinton facility. It succeeded in reducing emissions by 95 percent in one of the pilot sites. This innovation was then replicated in other sites within the Clinton facility, and, subsequently, extended to other Lilly facilities. Apart from the fix-it team, the role of the "SARA group" (in-company task force to implement Superfund Amendments and Reauthorization Act) was critical in organizing Lilly's attempt to reduce the releases of TRI chemicals. John Wilkins and Richard Lattimer, both of Corporate Environmental Affairs, played critical roles in organizing monthly meetings for tracking progress and intra-Lilly pollution-prevention conferences. As a result, there was a fair degree of employee involvement in programs for reducing emissions of TRI chemicals.

Methylene chloride continued to be emitted, albeit in smaller quantities. To reduce these emissions further, Lilly invested about $80 million in its two facilities at Clinton and Tippecanoe for installing state-of-the-art Regenerative Thermal Oxidizers (RTOs) or Fume Incinerators. All fumes from point sources, especially those containing volatile organic chemicals such as toluene and methylene chloride, were centrally collected and incinerated in these RTOs. Since installing the RTOs involved significant Type 2 expenditures that were not subjected to capital budgeting, they were initially opposed. However, policy supporters, especially Edward Smithwick, Vice-President of Bulk Manufacturing, and Richard Eisenberg, Executive Director of Corporate Environmental Affairs, emphasized the need for proactive measures, even though their benefits were non-quantifiable. The proposal for installing the RTOs was initiated in 1988. Due to initial opposition, the RTOs were finally installed in the Clinton facility in 1992 and in the Tippecanoe facility in 1993. Lilly also attempted source reduction by identifying non-toxic substitutes of methylene chloride. This involved employing a new management system known as NPERT (New Product Environmental Requirements Tracking). NPERT requires identifying environmental challenges in the early stages of process development of new drugs. Lilly employed NPERT in developing Raloxifene, a drug which can potentially treat and prevent osteoporosis. The Raloxifene team successfully eliminated methylene chloride and replaced it with a less volatile solvent that was easier to contain within the plant simply by altering the process chemistry. This team also eliminated/reduced other solvents (specifically polyphosphoric acid) from the manufacturing process resulting in a 74 percent reduction in total solvent use compared to the previous process. As shown in table 4.7, consequent to the efforts of the fix-it team, the new RTOs, and the NPERT process, Lilly reduced its releases of 33/50 chemicals by 81 percent between 1988 and 1995, well exceeding EPA's target of 50 percent reduction. The releases of methylene chloride, the key 33/50 chemical,

Table 4.7. *Lilly's performance on the 33/50 program*

	Total releases (in pounds)	Index versus baseline year 1988
1988	575,000	100
1991	450,000	69
1992	780,000	136 (EPA target: 67)*
1993	500,000	87
1994	100,000	17
1995	110,000	19 (EPA target: 50)

Note:
* This increase in releases was due to launching of new products requiring 33/50 chemicals for their manufacturing.
Source: Estimated from Lilly (1992b, 1995c).

were reduced by 77 percent by 1995 compared to 1988 levels. This team's efforts and the NPERT process were awarded the 1995 Indiana Governor's Award for Excellence in Pollution Prevention.

Baxter also adopted beyond-compliance policies in response to TRI. As reference, Baxter used at least eight of the seventeen 33/50 chemicals. These were: chloroform; chromium and chromium compounds; methylene chloride; methyl ethyl ketone; nickel and nickel compounds; toluene; 1,1,1,-trichloroethane; and trichloroethylene. Of these eight chemicals methyl ethyl ketone and 1,1,1,-trichloroethane were used in significant quantities (Baxter 1991a). As discussed previously, the TRI database has given external stakeholders access to information on facility-wise releases of toxic chemicals. Consequently, during 1989–1991 Baxter faced considerable adverse publicity in many states and counties where its facilities were located. This prompted some managers to suggest aggressive programs for proactively reducing the releases of TRI chemicals. A technical team within Baxter headed by Vice-President, Technical, Charles Kohlmeyer was established for evaluating the technical feasibility of reducing emissions of key TRI chemicals. A number of task-forces were set up to focus on individual chemicals. For example, Kohlmeyer himself chaired the task force on ethylene chloride and methanol and Rob Currie chaired the task force on trichloroethane. Consequently, Baxter has spent over $10 million on projects for reducing the releases of TRI and other chemicals.

These proactive steps for reducing the releases of TRI chemicals had significant *ex post* payoffs. In 1991, when the EPA invited Baxter to join the 33/50 program, Baxter already had on-going programs for meeting most of the 33/50 goals. On May 14, 1991, Vernon Loucks, Baxter's chief

Table 4.8. *Baxter's performance on the 33/50 program*

	Total releases (in pounds)	Index versus baseline year 1988
1988	873,000	100
1992	86,000	10 (EPA target: 67)
1994	42,000	5
1995	37,000	4 (EPA target: 50)

Source: Baxter (1996a).

executive, wrote to the EPA on Baxter's commitment to joining the 33/50 program. In his letter, Loucks committed to even higher standards than were required by the 33/50 program. Instead of a 50 percent reduction in releases by 1995, Baxter set a target of 80 percent reductions. In addition, Baxter established an 80 percent reduction goal based on equivalent product output for CFCs and 189 other chemicals covered by the Clean Air Amendments (Baxter 1991b). Importantly, Baxter set these *minimum* goals for facilities all over the world; facilities and divisions could set even more aggressive goals for themselves.

Baxter followed a three-stage process to meet the 33/50 program goals. It accorded source reduction the highest priority. If source reduction was infeasible, recycling was to be considered. If recycling was also infeasible, treating chemicals (and converting them to harmless compounds) was to be the route for meeting the 33/50 program goals. However, treatment was to be considered only as an interim measure; corporate research and development teams continued efforts to identify alternative processes or materials.

As shown in table 4.8, Baxter reduced its releases of the 33/50 chemicals by 96 percent: from 900,000 pounds in 1988 to about 37,000 pounds in 1995. This was a significant achievement because during this time period company-wide production increased by 20 percent.

Some of the key projects for implementing the 33/50 program are described below:

(1) Bentley Laboratories in Irvine, California, develops, manufactures, and markets disposable medical devices used for blood oxygenation during cardiopulmonary surgery. This facility reduced the emissions of 1,1,1-trichloroethane in two stages. In stage one, emissions were reduced by over 80 percent by 1990 by modifying process leach tanks and instituting operational changes. Initially, this facility resisted undertaking capital expenditure of about $50,000 for this task. However, persistence and persuasion by some managers, especially

Rob Currie, resulted in implementation of these initiatives. In stage two, by adopting a newly developed proprietary technology, which eliminated using 1,1,1-trichloroethane, these emissions were reduced to zero by 1991.

(2) The V. Mueller division manufactures a complete line of general and specialty surgical instruments, a line of endoscopic products for laparoscopic surgery, and the Genesis Container system. Mueller's St. Louis, Missouri facility totally eliminated the releases of 1,1,1-trichloroethane (8,000 pounds per annum in 1988) by installing an aqueous-based cleaning system.

(3) The Mannford facility in Oklahoma which manufactures a broad line of surgical and medical products such as disposable/reusable gowns, disposable drapes, and surgical packs, reduced its emission of methyl ethyl ketone by 60 percent between 1988 and 1994. This was achieved by modifying manufacturing process and reducing fugitive losses.

Although not required under the TRI or the 33/50 programs, Baxter's international facilities also significantly reduced their toxic emissions. Some examples:

(1) The Euromedical facility in Malaysia reduced methylene chloride emissions by over 100,000 pounds with 1988 as the baseline. This was achieved through changing material handling and modifying equipment.

(2) Methyl ethyl ketone emissions in the various maquiladora facilities located in Mexico near the US border were reduced by over 90,000 pounds with 1988 as the baseline by operational changes and tank modifications.

Analysis

TRI and 33/50 programs required Baxter and Lilly to spend millions of dollars in installing new equipment. Similar to intra-firm dynamics for investing in underground tanks, power-based explanations are inadequate since there was little dissent or bickering over the policies on TRI and 33/50. The reason is that these programs did not create "losers" who would have incentives to resolutely oppose them. There was indeed some initial opposition. However, after initial stalling, policy skeptics were persuaded to revise their assessments of the long-term benefits and costs of Type 2 expenditures on these programs. The relentless campaign by policy supporters played a significant role. In particular, they emphasized the critical need to remain in the good books of the EPA, an important external institution, particularly by adopting the 33/50 program. Thus, leadership-based explanations best explain the policy responses of these firms.

How did policy supporters convince policy skeptics? As indicated previously, the initial TRI reports generated adverse publicity for many firms. In 1988, Lilly was identified as one of the top ten emitters of TRI chemicals in the state of Indiana. Baxter was listed as a leading polluter in many states and counties (appendix 5.1). Further, *even when some firms reduced their releases of TRI chemicals, they continued to be identified as emitters of toxic materials*. For example, The *Arkansas Gazette*, in its eye-catching caption "Toxin level drops by half in 1 year, study says," reported that:

The latest figures from the federal Toxics Release Inventory show that Arkansas industries discharge less than half as many toxic chemicals in 1990 as in 1989. . . . Another example of dramatic reduction was Baxter Healthcare Corp. which reduced its total toxic emissions from 609,600 pounds in 1989 to 432,308 pounds in 1990. (1992: Arkansas Page)

Even complementary news items, such as the aforementioned, reinforce an image that firms continue to be significant emitters of toxic chemicals. For policy supporters, the message was clear: firms need to carefully protect their reputations because negative images tend to be sticky. Clearly, adverse media coverage created pressures on both Baxter and Lilly from their external stakeholders as well as employees for reducing releases of TRI chemicals.

Policy supporters further argued that the TRI program was a strategic step by the EPA to prepare the ground for stringent regulations. They suggested that the 33/50 program was a prototype for new laws if "voluntary" compliance from industry was not forthcoming. They highlighted the letter written by the EPA's Administrator, William K. Reilly to Vernon Loucks, the Chief Executive Officer of Baxter, noting that:

The American public has made clear that they expect nothing less than dramatic reductions in toxic chemicals releases. The challenge before EPA and industrial leaders is this: how do we bring such reductions about? One way is by the conventional command-and-control option which has been the Agency's mainstay for the past twenty years. But I believe there is another, more fruitful path that we can follow which is faster, and *without the detailed direction which is likely to be demanded by the public if voluntary efforts are not fruitful* (Baxter 1991a; italics mine).

For policy supporters, Reilly's message was very clear: either industry voluntarily commits to such reduction, or citizens will demand that the government intervene. Such interventions, typically in the command-control mode, will leave little autonomy for firms on issues such as which chemicals to reduce, which technology to employ, and the time frame of implementation. Assuming that Riley's threat was credible, what incentives did individual firms have to actually join the 33/50 program or to voluntarily reduce the release of TRI chemicals? Why could individual firms not free-ride on the efforts of other firms? How did the institutional design of the 33/50 program and the TRI reporting requirements mitigate collective action dilemmas?

The TRI reporting requirements and the 33/50 program created two categories of benefits. First, citizens benefitted from having a cleaner environment. Since these benefits are non-excludable, firms producing such benefits cannot force citizens to pay for them. As a result, firms have little incentive to provide these benefits. Second, these programs create goodwill benefits for firms among regulators and citizens. These programs also raise the morale of employees. Such goodwill benefits accruing to firms are excludable since it is possible to identify firms that have invested in reducing the releases of TRI chemicals or have joined the 33/50 program. Citizen groups and journalists often monitor firms' performance on reducing TRI emissions, and advertise them. The EPA also enhances the excludability of such benefits by regular media reports on 33/50 success stories or trends in releases of TRI chemicals.

What were the strategies adopted by policy supporters to convince their skeptical colleagues of the benefits of such Type 2 programs? As discussed previously, the first TRI report was sent to the EPA in July 1988. In the process of quantifying releases of TRI chemicals, policy supporters in Lilly realized the potential for bad press once this information became public knowledge. As early as 1988, they suggested that Lilly should be prepared to invest millions of dollars to proactively reduce its emissions of methylene chloride. At first, there was resistance to this proposal since its benefits were not quantifiable. Skeptical managers argued that since Lilly adheres to existing emission laws, there was no need for further reductions at a substantial cost. However, the supporters pointed out that consequent to the TRI reports, Lilly had been identified as one of Indiana's leading emitters of TRI chemicals. The Clinton and Tippecanoe facilities had been ranked fourth and fifth nationally as emitters of methylene chloride. In addition, it is necessary for Lilly to remain on good terms with the EPA and proactive reductions in emissions of TRI chemicals would facilitate this.

Edward Smithwick, Vice President, Bulk Manufacturing, and Richard Eisenberg, Executive Director of Corporate Environmental Affairs, in particular, were key figures in persuading their skeptical colleagues of the long-term benefits, even though not quantifiable, of proactively reducing TRI emissions. The advocacy by Smithwick was important since the expenditures on reducing TRI emissions were borne by facilities. Since every facility is a profit center, one expects facility managers and their organizational superiors (such as Smithwick) not to support programs that reduce quantifiable profits. Once the TRI chemical reduction program was in place, 33/50 targets could be achieved easily. Consequently, in 1991, when the EPA invited Lilly to join the 33/50 program, policy supporters met with only marginal opposition. The skeptics saw little value of joining yet another EPA-sponsored program. However, policy supporters pointed out that since Eisenberg served on the Federal Advisory Committee on the Clean Air Act, it was difficult for Lilly not to join the 33/50 program. Further, they pointed out that the 33/50 program would require only minor changes in the internal organization. This of course reassured the skeptics that the organizational status-quo would not be upset which may have inflicted "losses" on them.

Baxter also invested in excess of $10 million for reducing its releases of the TRI and other toxic chemicals. Policy supporters adopted an interesting strategy by making this project a part of Baxter's overall objective of establishing a state-of-the-art environmental program. Since the policy skeptics could not exercise a line-item veto, and they were in favor of other items of the state-of-the-art program, they had few incentives to derail the whole program.

As discussed in chapter 3, Baxter's environmental program faced rough times in the mid 1980s primarily resulting from Baxter's merger with the American Hospital Supply Corporation. In 1989, some of Baxter's senior managers realized the long-term costs of having reactive environmental policies. This coalition of senior managers, specifically, William Blackburn, Charles Kohlmeyer, and G. Marshall Abbey (and, subsequently, Arthur Staubitz), argued for establishing a state-of-the-art environmental program within Baxter. They eventually managed to convince Vernon Loucks, Baxter's Chief Executive, of its long-term payoff. Vernon Louck had presided over the controversial merger of Baxter with AHSC and was therefore inclined to take necessary steps that would reduce the potential flak on environmental issues. In the process of operationalizing this ambitious mandate, policy supporters included programs for reducing the releases of TRI chemicals. They very astutely linked these programs to reductions in the emissions of CFCs, chemicals that had been identified as main causes of the Ozone Hole and whose use and production had been regulated under the Montreal Protocol of 1987. Further, the TRI reduction program was subsumed under a broader program for waste reduction and pollution prevention. These were very clever political strategies where by TRI and 33/50 programs were bundled with popular and established policies.

Nevertheless, there was opposition from facility managers on two counts. First, they were hesitant to commit to ambitious and still undefined state-of-the-art goals. Second, since facilities were profit centers, they bore the costs of such programs. They were therefore reluctant to bear costs for projects that imparted meager excludable and quantifiable benefits to their facilities. Further, they feared intrusion by the Corporate department in their environmental policies under the garb of these programs.

Policy supporters read this opposition carefully. They decided to induce acceptance by addressing the concerns of policy skeptics through persuasion and not imposition. First, they finalized the program goals only after actively consulting with the facility and divisional managers. Thus, only after skeptical managers had been convinced of the technical feasibility and financial support for these programs did the policy supporters propose the program goals to the top management.

Second, policy supporters consciously decentralized the implementation of the various pollution prevention and waste minimization programs. This encouraged facility managers to take the lead in such programs, to identify solutions, and, importantly, to take the credit for their success. Thus, the decentralized structure aligned incentives of the corporate, divisional, and facility managers, thereby enthusing the hitherto skeptics to ensure that the programs were adopted.

The policy supporters regularly publicized the progress made by facilities and contributions of specific managers towards reducing the releases of TRI chemicals, 33/50 chemicals in particular (for example, Baxter, 1994a). Baxter's annual environmental conferences, often attended by about 200 managers, were extremely valuable in providing in-company visibility to high performers. Also, Baxter's bi-monthly in-house magazine *PACE*, circulated to all its 60,000 employees consciously highlighted the achievements of such programs. This created incentives for the facility managers to enthusiastically adopt this program since they did not want to be identified as Baxter's "bad guys." In his letter to the EPA, Rob Currie, Baxter's 33/50 coordinator noted that:

Baxter has found that communication and awareness are key to implementing successful pollution prevention and waste minimization programs. As a result, several vehicles to foster communication are used to bolster facility programs . . . (Currie 1995)

To conclude, leadership-based theories best explain why and how Baxter and Lilly invested millions of dollars in reducing releases of TRI chemicals. The TRI and 33/50 programs were initially opposed on account of their significant financial commitments and their potential to bind facility managers to relatively undefined goals. Policy supporters did not trample upon this opposition, as power-based theories would predict.

Instead, they sought to overcome it by bundling these programs with other widely popular policies, and, importantly, by genuinely attempting to address the concerns of the facility managers. They also emphasized the need to remain in the good books of the external stakeholders, especially the EPA. Since the role of key managers was critical in inducing this consensus, the policy adoption is best understood by employing leadership-based theory.

3   Responsible Care[5]

This section examines Eli Lilly's response to the Chemical Manufacturers Association's Responsible Care program. Specifically, why did Lilly first resist and then enthusiastically adopt this program? Given the nature of its product portfolio (with one of the 200 facilities involved in manufacturing chemicals), Baxter has virtually no involvement with the CMA, and consequently with Responsible Care. Hence, this section focuses on Lilly only. My conclusion is that power-based theories best explain Lilly's initial reluctance, and leadership-based theories its subsequent enthusiastic adoption of Responsible Care.

Founded in 1872, the Chemical Manufacturers Association (CMA) is one of the oldest trade and industry associations in the US. Its 200 members together account for about 90 percent of US industrial capacity in basic chemicals (CMA 1996a). Responsible Care is CMA's flagship program for improving the chemical industry's performance in the areas of health, safety, and the environment. Versions of Responsible Care have been proposed by CMAs in other countries as well. The US CMA mandates that all its member adopt Responsible Care. Members, however, have the autonomy on the pace of its implementation, and, conceivably, members may not implement it at all.[6]

---

[5] This section draws on Prakash (1999a).

[6] Responsible Care was launched first in Canada and brought to the US and other countries only subsequently. The subsidiaries of US multinational enterprises (MNEs) such as Dow-Canada were early adopters of Responsible Care and played a significant role in establishing the legitimacy of this program in Canada within chemical firms. Many of Responsible Care's ideas were developed by Dow-Canada in response to the accident in its Sarina facility. Some individuals such as Dave Buzelli (Dow-Canada) and Robert Kennedy (Union Carbide) played key roles in popularizing Responsible Care in their firms, their US parent, and the Canadian/US CMA. It is said that during a review of Canadian operations, Robert Kennedy, Chief Executive Officer of Union Carbide, was informed of this initiative. He quickly gauged the potential benefits of having this as an industry-wide program in the US and passed this information on to the US CMA's Public Perception Committee. The committee bought into Kennedy's suggestion and consequently, the US CMA launched Responsible Care in 1988 (*Chemical and Engineering News* 1992).

Responsible Care was launched in the backdrop of economic success but stakeholder distrust of the US chemical industry. The industry's economic performance was impressive: it was ahead of most others in exports, research and development expenditures, and wages paid to its manufacturing workers (*Chemical and Engineering News* 1992). However, its credibility with regulators and other external stakeholders on environment, health, and safety issues was plummeting. A series of major chemical accidents, including the Bhopal disaster in 1984, reinforced a perception that the chemical industry cannot conduct its operations without harming human health and damaging the environment. Consequently, there were demands for governmental intervention. Industry leaders feared that such high levels of policy activism and the accompanying uncertainty would undermine investors' confidence in the long-term prospects of the chemical industry and hurt its stock prices. Further, they argued that the ever-increasing compliance requirements would divert scarce resources from research and development, and eventually hurt the industry's international competitiveness.

In 1983, one year prior to the Bhopal tragedy, the CMA had developed a statement of principles on how the chemical industry should conduct its business and relate to its stakeholders. This statement eventually became the basis for developing the Ten Guiding Principles of Responsible Care (see below). However, the scale of the Bhopal disaster altered the character of discourse on how the industry ought to be regulated; immediate and drastic actions were demanded. To introduce public accountability of its activities, the CMA proposed a voluntary program, Community Awareness and Emergency Response (CAER), in 1985. Eventually CAER became one of the Six Codes of Responsible Care (see below). Since many environmental groups saw CAER as a public relations gimmick, in late 1985, the CMA formed the Public Perception Committee composed of top industry executives. This was a precursor to the Executive Leadership Groups, an important feature of Responsible Care. The Public Perception Committee recommended that the CMA launch Responsible Care *(Chemical and Engineering News* 1992).

Responsible Care was launched in the US in 1988. Three categories of actors have subscribed to Responsible Care: (1) members of the CMA; (2) partner companies, particularly those in the transportation sector: railroads, trucking, and barge (these firms are not members of the CMA); (3) partner associations such as the state chemical industry councils and the national associations of firms that deal with chemicals.

This program has the following features (*Chemical and Engineering News* 1993b; CMA, 1996a,b):

(1) Ten *Guideline Principles* spelling out responsibilities of CMA member firms.

(2) Six *Codes of Conduct* that identify more than one hundred specific management practices. These codes seek to establish management systems in manufacturing, distribution, and transportation (details in appendix 4.2).

(3) A fifteen member *Public Advisory Panel* consisting of non-CMA members to guide Responsible Care. This panel is expected to sensitize the CMA to public concerns and give input on developing programs that better address these concerns.

(4) A *requirement* that all the member firms will adopt Responsible Care. Firms are not required to implement all the six codes immediately; they can chart their own time frame for implementation.

(5) Member firms must *annually evaluate* their progress on implementing the six codes of Responsible Care. This evaluation should be shared with the CMA.

(6) *Executive Leadership Groups* of senior industry representatives to periodically share their experiences on implementing Responsible Care and to identify areas requiring assistance from the CMA.

Critics are skeptical of the impact of Responsible Care on the culture and functioning of the chemical industry. They view it as an attempt by industry to preempt stringent legislations (Chatterjee and Finger 1994). Fred Miller of Friends of Earth views Responsible Care as "the velvet glove around the iron fist . . . industry has a heavy burden of proof to show that it is just not a PR gimmick" (*Chemical and Engineering News* 1992: 39). Further, many dispute the extent of change in the culture and practices of the chemical industry, and the speed with which they have been accomplished.

Further, since CMA membership is voluntary, a threat of expulsion for not implementing Responsible Care is not credible. This is also tacitly accepted by Jon Holzman, CMA's Vice President for Communication who maintains that the CMA does not intend to expel members for not implementing Responsible Care. Rather, CMA's strategy is to encourage and equip members to implement it. According to Holzman, the CMA follows a three-step process for monitoring progress on Responsible Care and correcting unsatisfactory performance. First, the CMA talks directly to the facility managers (since progress is monitored at the facility level) and gives them deadlines for correcting implementation shortfalls. If this fails, then the CMA contacts senior managers at the firm's headquarters and encourages them to pressure their facility managers to improve performance on Responsible Care. If this does not succeed, then conceivably a firm may lose its CMA membership (*Chemical and Engineering News* 1992).

Lilly: policy dynamics

When the CMA proposed Responsible Care in 1988, Lilly was hesitant in adopting it; or, if forced to adopt, an early internal assessment suggested that it would not implement all of the codes. In particular, some senior managers strongly objected to the code on community awareness and emergency response that required Lilly to develop community outreach programs. They argued that since Lilly was following all the applicable laws, there was little justification for developing such programs. Further, since the public has little technical knowledge of the manufacturing processes, information provided by Lilly could be misinterpreted. Consequently, community outreach programs would create confusion, thereby not serving their intended purpose. Due to the significant hierarchical power of policy skeptics, Lilly did not propose community outreach programs for a few years. Its environmental managers also kept a low profile in discussing this program with employees as well as with external stakeholders.

However, post 1993, Lilly has become a show-piece of successful implementation of Responsible Care. It has also become very active in the CMA on this subject: John Wilkins of Lilly's Corporate Environmental Affairs chairs the CMA's Indiana-Ohio-Kentucky chapter of Responsible Care. Currently, Lilly is implementing Responsible Care codes in the following sites: Carolina (three facilities), Puerto Rico, Clinton (Indiana), Indianapolis (Bulk, Pharmaceutical, and Chemical Process Development, Indiana), Mayaguez (Puerto Rico), and Tippecanoe (Indiana). Even though not required by the CMA, some of Lilly's foreign affiliates are also implementing their respective national versions of Responsible Care. These affiliates included Cosmopolis (Brazil), Kinsale (Ireland), and Speke (UK).Three examples of innovative Responsible Care programs are given below (Lilly 1994c):

(1) Lilly has teamed up with some Indianapolis-based firms for establishing community outreach programs. In addition, many Lilly facilities also undertake solo initiatives. For example, the Tippecanoe facility in Indiana has organized a community forum for discussing its environmental, health and safety performance. This forum meets periodically, typically twice a year. This facility also conducts an annual emergency drill with the county's police, fire, and health services. As a result, the Tippecanoe facility has brought in more transparency in

its decision-making processes and has generated confidence within the local community about its ability to deal with industrial accidents.

(2) Previously, I described how Lilly networked with other Indianapolis-based firms to establish community outreach programs. Lilly, together with these firms, has also established a logistics system for chemical shipments. For example, in the early 1990s, the state of Indiana did not have the necessary statistics for planning shipping operations. These firms, together with their transporters and city/state officials, created a task force for gathering such statistics. This enabled these firms, including Lilly, to schedule shipments in a cost-effective manner.

(3) Lilly has developed a computer-aided logistics system that enables its traffic managers to route shipments through the safest routes. It also closely scrutinizes the safety records and policies of its transport carriers. Further, it has an emergency action plan in case an accident should occur.

As described previously, the CMA recommends that firms annually assess their facility-wise implementation of Responsible Care. As shown in tables 4.9 and 4.10, Lilly evaluates the facility-wise progress on all six codes of Responsible Care. It has invested considerable resources in creating a six-point scale, and auditing facilities on their progress on all six codes. In fact Lilly conducts multiple Responsible Care audits of its facilities every year and reports the annual average scores for every facility in its Environmental Performance Reports. Lilly has set a goal that every facility should score at least 5.0 on each of the six codes. Since progress on Responsible Care is an important input in the evaluation of the facility managers, they have significant incentives to implement this program. In addition, Lilly's Tippecanoe facility is implementing the CMA's pilot program – Management Systems Verification – where external auditors assess a facility's progress on Responsible Care (CMA 1998). This again reinforces the fact that Lilly is currently in the forefront of implementing Responsible Care.

Analysis

Power-based theories are helpful in explaining Lilly's initial reluctance and leadership-based theories in explaining its subsequent enthusiastic implementation of Responsible Care. When the CMA proposed Responsible Care in 1988, some senior managers were hesitant in adopting it due to the code on community awareness and Emergency Response

Table 4.9. *Lilly's internal evaluation of Responsible Care*

	1992	1993	1995*
Community awareness	3.7	3.7	4.5
Emergency response	4.5	4.5	4.7
Pollution prevention	3.6	3.7	4.4
Process safety	3.6	4.0	4.0
Distribution	2.7	3.3	n.a.
Health and safety	4.3	4.7	4.5
Product stewardship	2.0	4.4	n.a.

Notes:
Scale
1 = no action has been taken
2 = evaluating existing experience
3 = developing plans
4 = implementing plans
5 = plans have been implemented and are in place
6 = reassessing the existing plan
* These scores represent the mean of scores of six Responsible Care facilities.
Source: Adapted from Lilly (1992b, 1995c).

that required developing community outreach programs. They argued that communities living in the vicinity of any facility often have little technical knowledge for appreciating how manufacturing processes work. Hence any sharing of technical information may lead such communities to make unwarranted conclusions about the safety aspects and environmental impacts of a facility's operations. As discussed previously, establishing community outreach programs was opposed by many leaders in the chemical industry as well. As Simmons and Wynne observe:

Fundamental to the identity of the chemicals sector is its sense of being a science-based industry. This is deeply ingrained in the industry's culture and belief about the validity and authority of science frame its view of outside groups. These beliefs are reflected in the argument that has been made to legitimate the *industry's claim to self-regulation – that its unmatched knowledge and expertise make the industry's own experts the people best suited to audit and regulate the environmental effects of its activities.* (1993: 218; italics mine)

Thus, even though Lilly had to formally adopt Responsible Care, it kept a low key approach, especially on implementing the code on community awareness and emergency response. In 1993, after about five years of formally adopting Responsible Care, when one of the main opponents (a senior manager) of this program retired, policy supporters spotted a window of opportunity. They were in a better position to convince the

Table 4.10. *Responsible Care: implementation status*

Site	1995 average*	1991 average*	Percent of practices in place in 1995
Carolina	4.1	2.7	28
Clinton	5.0	3.3	73
Indianapolis	4.7	2.7	61
Mayaguez	4.7	2.7	62
Tippecanoe	5.1	3.3	66
Total	4.7	2.9	58

Notes:
Scale
1 = No action has been taken
2 = Evaluating existing experience
3 = Developing plans
4 = Implementing plans
5 = Plans have been implemented and are in place
6 = Reassessing the existing plan
* Since Lilly conducts multiple audits every year, average of all audits scores is being reported.
Source: Wilkins (1996).

remaining skeptics of the benefits of implementing Responsible Care. Particularly, policy supporters highlighted the tremendous benefits, even though non-quantifiable, of establishing community outreach programs in terms of generating goodwill within the CMA and among local communities, regulators, and environmental groups.

Why and how did policy supporters succeed in convincing policy skeptics of the benefits of being in the forefront of implementing Responsible Care? Specifically, what benefits accrue to Lilly for maintaining a high profile on this subject within the CMA? As I have discussed before, the CMA is a significant player in policy discussions on proposed laws and regulations or any issue that impacts the chemical industry. Consider these two examples. The EPA has proposed to increase the number of chemicals to be reported under the Toxic Release Inventory program. This has significant cost implications for the chemical industry since they will come under stakeholder pressure to reduce their emissions of the chemicals added to the TRI list. The CMA has challenged this proposal in a federal court on the grounds that the new additions do not constitute significant threats to public health and therefore should not be clubbed with the original TRI chemicals. It appears that eventually the EPA, the CMA, and other parties will compromise on this subject; the EPA will perhaps scale down the list of additions. Firms having a crucial say in the running of the CMA will have better chances of influencing the eventual outcome in their favor.

The second example pertains to the CMA's response to the public debate generated by the book, *Our Stolen Future* (Colborn 1996). This book suggests that certain chemicals may disrupt the normal functioning of endocrine glands, thereby leading to problems such as sterility. Consequently, many environmental groups have called for a ban on the use of such chemicals. Since evidence on this subject is inconclusive, the CMA has committed $850,000 for further research into the causes of endocrine disruption. Again, the main actors in the CMA will play a crucial role in shaping the CMA's response to this controversy that has bearing on the future of the chemical industry. These examples suggest that firms indeed have significant incentives to gain credibility within the CMA, thereby influencing its agenda. However, since such benefits are not quantifiable, policy supporters had to invest significant efforts to convince the skeptical managers of the potential payoffs of implementing Responsible Care.

One of the concerns of policy skeptics was that implementing Responsible Care would be very expensive. Since it is difficult to assess the adequate level of expenditure on programs such as community outreach, Lilly would have a propensity to overspend. In response to these concerns, Responsible Care supporters developed innovative methods for controlling community outreach program costs. Eli Lilly established a network with other Indianapolis-based firms such as Reilly Industry, National Starch, and Dow Elanco each having facilities in close physical proximity. These firms jointly undertake initiatives such as community outreach activities and conducting science demonstrations at local elementary schools. Thus these firms share costs as well as provide a superior program by pooling their talents (*Chemical and Engineering News* 1993a).

An important task for the policy supporters was to ensure wide-spread internal participation in developing and implementing Responsible Care. Specifically, there was a concern that Responsible Care should not be viewed as a corporate program over which facilities had little control. To meet this objective, a Responsible Care group was established with representatives from various facilities. This group meets once every two months to share experiences in implementing existing Responsible Care initiatives and developing new ones. Establishing this group has resulted in unprecedented enthusiasm among facility-level managers for implementing Responsible Care.

Policy skeptics also argued that there was little clarity about the expectations from facility managers on Responsible Care. This was a crucial issue since most of the Responsible Care programs were to be implemented at the facility level. Policy supporters handled this concern by proposing a six-point scale (tables 4.9 and 4.10) to assess progress on the various codes of Responsible Care. Further, they also established a system of regular internal audits, whereby facilities could have specific and timely feedback on their progress in implementing this program.

To conclude, an important learning from this case is that individuals matter, not only in implementing particular programs, but also in opposing them. Structurally advantaged individuals can ensure the implementation of unpopular policies or thwart the implementation of popular policies. Thus, power-based processes can work both ways: in having policies implemented and in having policies jettisoned. Power-based processes therefore explain Lilly's initial hesitation in implementing all codes of Responsible Care.

Leadership-based processes explain why Lilly successfully implemented all codes of Responsible Care, even though the benefits of such policies were not quantifiable. Once the key opponent of Responsible Care retired from Lilly and was not in a position to influence its environmental policymaking, policy supporters succeeded in generating consensus, especially on implementing community outreach programs. They convinced the skeptics that adopting all codes of Responsible Care would serve Lilly's long-term interests, even though such policies may not generate quantifiable profits. They highlighted that aggressive and visible adoption of Responsible Care would increase Lilly's credibility within the CMA, and with the regulators, environmental groups, and local communities. Credibility within the CMA was particularly important, given the CMA's preeminent position as the representative of the chemical industry in most environmental policy debates. Policy supporters won the trust of skeptics by innovating strategies for controlling costs and enhancing the quality of Responsible Care initiatives. They ensured that no significant group within the company perceives itself as a loser from implementing this program, and hence have incentives to oppose it. This was important as many Responsible Care programs, especially on community outreach, are quite revolutionary in that they expose managers involved in extremely technical tasks to public scrutiny. Thus, a consensus mode of implementation ensures that technical experts do not feel threatened by imposition of such programs, and, at the same time, the firm meets expectations of its external stakeholders in terms of sharing information on its internal functioning.

4   "Green products"

This section focuses exclusively on policy dynamics within Baxter on the marketing of green products. Green products are those that specifically aim to deliver enhanced environmental quality as one of their key product benefits (Coddington 1993). Given the nature of Baxter's product portfolio, it had opportunities to offer such products, and the presence of certain key managers within the firm ensured that such opportunities were tapped. Some products used vast quantities of packaging which imposed significant disposal costs on Baxter's customers.[7] Environmentalists were also concerned about the environmental implications of excessive packaging. Due to the vision of key managers, Baxter aggressively reduced packaging, redesigned its products, and satisfied latent customer need for environmentally friendly low-disposal cost products. These strategies were not justified by using formal investment criteria and yet they did not result in intra-organizational conflicts. Hence, leadership-based explanations best explain Baxter's decision to market green products.

Given Lilly's product portfolio, it does not have a business rationale, in the short or the long run, to manufacture and market green products. The reason is that environmentally friendly features in ethical or prescription drugs (Lilly's main line of business) do not motivate doctors to recommend them to their patients; the cost of treatment, efficacy, and lack of side-effects are the key motivators in brand choice. Thus, developing green ethical drugs or portraying its existing drugs as green provides meager quantifiable or non-quantifiable benefits to Lilly. It is important to reiterate that this book examines cases where firms have adopted policies that are perceived by managers to provide non-quantifiable benefits that accrue in the long-run. Thus, although such policies cannot be defended by employing capital-budgeting, managers believe that they are important and beneficial for the firms' survival and prosperity. This book *does not* make an argument that managers advocate policies that provide little benefits, quantifiable or non-quantifiable, to their firms.

The greening of firms can imply both the greening of their value-addition processes and of their products. An important question therefore is: how does greening influence the purchase decisions of customers? To

---

[7] Customer means the product's purchaser while consumer means the product's user. For a firm marketing final products, distribution chains are its customers, and individual actors purchasing from these outlets are its consumers. For industrial and intermediary products, where one firm sells to another, the term customer is more applicable. For example, if Baxter sells dialysis machines to health-care centers, these healthcare centers are Baxter's customers.

explore this subject, this book examines two dimensions of greening. First, verifiability of greening by customers or actors external to firms. Second, importance of greenness in influencing purchase decisions. Apropos the first dimension, customers often have little information on this subject, and, consequently, they find it difficult to factor in environmental aspects of firms' processes and products in their purchase decisions. Such information asymmetries therefore lead to market failures.[8] To mitigate market failures, policy interventions take place at various levels. Governments enact laws requiring firms to report on aspects of their environmental program. For example, under the Toxic Release Inventory Program, firms with manufacturing facilities in the US are required to report annual releases of specified toxic chemicals. Likewise, firms seek to mitigate market failures at the industry level by sponsoring industrial codes of environmental self-regulation such as the CMA's Responsible Care program. Firms also seek to mitigate information asymmetries at their level only. For example, they obtain environmental certifications such as ISO 14000 that are awarded by credible external auditors. All such measures, whether at the macro, industry, or firm level, seek to provide credible information on the greenness of firms and their products. Consequently, they help customers to factor in firms' greenness in their purchase decisions.

The second dimension I examine is whether customers consider greenness of firms an important factor in their purchase decisions. Customers base their purchase decisions on a variety of factors such as price, performance, availability, and reliability of products as well as the corporate image of firms that are manufacturing and marketing them. Consequently, the level of greenness of firms and their products is one of the many factors influencing the purchase decision; customers may or may not prioritize it over other factors.

As discussed in table 4.11 below, based on the two dimensions – verifiability of greenness and importance of greenness – we can identify four categories of situations.

Cell 1 represents market situations in which customers not only find it difficult to verify firms' claims on greenness, they also do not attach much importance to them. For example, customers will find it difficult to verify whether a firm has replaced its underground storage tanks with expensive

---

[8] Based on degrees of information asymmetries, goods can be classified into three categories: search goods, experience goods, and post-experience goods. Consumers can determine the characteristics of a: (1) search good prior to purchasing it; (2) experience good after purchasing it; and (3) post-experience good only with a lag after consuming it. Thus, information asymmetries (and concomitant market failures) are most severe in post-experience goods and least, if at all, in search goods (Weimer and Vining 1992).

Table 4.11. *Impact of greening on customers*

	Customers cannot verify greenness of firms' policies or of their products	Customers can verify greenness of firms' policies or of their products
Environmental issues are unimportant in influencing customers' purchase decisions	Beyond-compliance policies such as removing underground tanks do not influence brand choice of customers of anti-depressants or heart valves (Cell 1)	Adopting ISO 14000 does not influence brand choice of customers of anti-depressants or heart valves that are located in the US. (Cell 4)
Environmental issues are important in influencing customers' purchase decisions	These situations are unlikely to occur. If firms recognize that customers value greenness, then most of them can be expected to provide credible information on it (Cell 2)	ISO 14000 certification is required for selling in some European countries. Many US-based hospitals buying medical products from Baxter value green attributes of such products (Cell 3)

above-the-ground tanks, whether it conducts internal environmental audits, or whether heart valves are bio-degradable. Importantly, even if such claims are verifiable, the adoption or non-adoption of such green policies would not influence some customers' brand choice. Consider an individual wanting to purchase an anti-depressant. It is unlikely that this individual would purchase Prozac, an anti-depressant produced and manufactured by Lilly, simply because Lilly has removed underground storage tanks from its facilities and replaced them with expensive above-the-ground tanks, or because Prozac is bio-degradable, or because it is packaged in recycled material. This individual probably has little access to information on Lilly's environmental policies. And, even if the customer has such access, this knowledge may not significantly affect his/her purchase decision. For an anti-depressant, the customer may focus pre-dominantly on Prozac's performance and cost of treatment, giving little importance to Lilly's environmental policies or to Prozac's environmental impact. Similarly, customers wanting to purchase heart-valves will prob-ably not buy them from Baxter just because they are bio-degradable, or because Baxter invited Arthur D. Little (ADL) to help it define state-of-the-art environmental standards, and then audit its environmental pro-grams against them.

Cell 4 represents market situations in which customers can verify levels of greenness, but they do not consider it important in choosing brands. For example, firms may have the ISO 14000 certification, be members of Responsible Care, or subscribe to the EPA's 33/50 program. All these are examples where the greening of firms can be verified. However, this greening may be unimportant for customers wishing to purchase anti-depressants or heart valves.

If customers do not attach importance to the greenness of firms' policies or products (Cells 1 and 4), supporters of such environmental policies within firms probably do not justify them in terms of their impact on customer goodwill. As discussed in previous sections of this chapter (underground tanks, the 33/30 program, and Responsible Care), the adoption of Type 2 policies is best explained by understanding how policy supporters managed to convince the organization of the long-term benefits of these policies. These benefits include reducing environmental risks, and having the goodwill of regulators, local communities, and industry-level organizations.

Cell 2 represents market situations in which customers consider the greenness of firms or their products important but they have little access to verifiable information on this subject. Consequently, customers have to rely on claims made by firms or on media reports. Such situations are unlikely because if firms recognize that customers value greenness, they can be expected to make efforts to provide information on this subject. If anything, firms are often perceived to be over-stating the achievements of their environmental programs.

Cell 3 represents those situations in which customers are able to verify levels of greenness and they also consider greenness important in influencing their brand choice. For example, purchase decisions of some European customers are critically influenced by whether or not a supplier has the ISO 14000 certification. Or, some customers may only purchase products having a green dot.

Note that greenness can be certified at the firm level (ISO 14000) as well as at the product/brand level (green dot). This distinction becomes important when thinking about business strategy. Do customers buy products based on brand's attributes only or are their purchase decisions also significantly affected by a company's image and philosophy (Charter 1992)? If purchase decisions are significantly affected by a company's image, firms have incentives to invest in firm-level programs and then become certified. As discussed previously, for certain categories of customers, firm-level certification is indeed very significant.

However, if customers buy brands, and not a firm's philosophy, firms have incentives to green their products, to certify, and to communicate it to their customers.[9] Thus, customers can examine the attributes of such products, make inferences about their environmental impacts, and factor them into their decisions on brand choice. This is the focus of this case: how do firms deal with beyond-compliance policies that influence the attributes of their products, given that these attributes are valued by their customers?

Baxter and Lilly: response

Lilly manufactures and markets ethical pharmaceuticals. Some of Lilly's products such as insulin are over-the-counter (OTC). However, since patients often purchase these products on doctors' recommendations, such OTCs are *de facto* ethical drugs. For ethical drugs, the purchaser is not the decision maker regarding brand choice: patients are the purchasers and their doctors are the decision makers. Research suggests that among various attributes of an ethical drug, doctors consider its efficacy most important; after all their professional reputation depends on the drug's performance.[10] One could argue that other things being equal, the "greenness" of the drug may become important. That is, if a doctor is asked to choose between two ethical drugs that are equal in all aspects such as efficacy, cost, and reputation of the manufacturer, then this doctor may prefer the greener drug. However, such other things are seldom equal. Even though generic drugs such as basic antibiotics manufactured by different firms may be equivalent on most major attributes, by employing various tools of marketing, firms often succeed in differentiating their brand from the rest.

Further, since the health-care industry is highly research intensive, new drugs claiming better efficacy are being continually introduced. As a result, it is difficult to imagine that doctors will face similar situations on brand choice for ethical drugs in any major disease category. Of course, one can argue that doctors may be willing to prescribe greener drugs that

---

[9] Companies such as Procter and Gamble, General Mills, and Colgate focus on communicating the benefits of their brands. I am not arguing that such brand-focused firms ignore their corporate image. They do not. However, such firms focus most of their advertizing on talking about their specific products. For example, most of Procter and Gamble's advertizing focuses on the superior performance of Tide detergent, the freshness of Ivory soap, or the beauty-enhancing effect of Oil of Ulay. Most consumers probably do not link all these brands to Procter and Gamble. Of course, some other firms focus on corporate advertizing or have a generic brand name for their various products. In particular, Japanese companies have generic brand names. Thus, we have a SONY television, SONY VCR, and SONY Walkman.

[10] I have similar findings from my previous career at Procter and Gamble as a brand manager of the Vicks range of OTC drugs.

are only marginally less efficacious (or marginally more expensive) than non-green drugs, as long as greener drugs are efficacious enough (or cheap enough) for treating patients. Such possibilities certainly exist for only basic drugs such as basic antibiotics or sulpha drugs, and not in Lilly's product categories.

I am not arguing that firms' environmental policies do not impact doctors' brand choice for ethical drugs. They certainly do. Suppose there are extensive media reports on pollution caused by a pharmaceutical firm or adverse environmental impacts from its products. Surely doctors may begin to have negative opinions about this firm and its brands. However, adoption of beyond-compliance policies or greening products will probably not persuade doctors to prescribe the firm's brands. To employ Herzberg's (1966) terminology, for doctors, a lack of negative publicity about the environmental policies of a firm, or the environmental impact of an ethical drug is a "hygiene" factor.[11] However, the presence of environmentally friendly features in ethical drugs or the adoption of beyond-compliance policies by a firm is often not a "motivator" for recommending the brands of ethical drugs manufactured by this firm. Factors such as the cost of treatment, efficacy, and lack of side effects are the motivators' for doctors in brand choice. On this count, Lilly has little reason to market green ethical drugs or to portray its existing drugs as green.

In contrast to Lilly, Baxter markets green products: medical supplies that are purchased by hospitals and other health-care suppliers. Some of Baxter's green products are briefly described below (Baxter 1994d; 1996b).[12]

(1) Access: Baxter and Waste Management, Inc. have formed a strategic alliance to offer Baxter's corporate customers solutions to all their environmental issues. Waste Management, Inc. is the world's largest waste disposal company offering consulting services on waste reduction, recycling, treatment and disposal of waste, etc.

(2) ValueLink: Baxter repackages bulk products to customer

---

[11] Extending Maslow's (1943) theory, Herzberg develops a theory of work motivation. His study focuses on two work-related factors: those that turned people on and those that turned them off. He labels the former 'motivators'; these factors pertain to the job content such as levels of creativity and recognition. He labels the latter 'hygiene' factors; their presence only prevents dissatisfaction among employees. These factors are related to the job context such as company policies, working conditions, and salary.

[12] Baxter sought to address customers' environmental concerns in other ways as well. In 1995, it established a Customer Waste Council (CWC), a multi-divisional and cross-functional initiative to better respond to customers' environmental concerns. As of 1997, the CWC consists of five teams whose objectives include evaluating the latest technologies in reusable shipping containers; investigating what environmental information consumers are requesting, why they are making these requests, and how Baxter can fulfil them (Baxter 1996b).

specifications, delivers them daily to the patient floors using reusable totes, and recycles their bulk-packaging. Consequently, this eliminates bulk-packaging from customers' waste streams.

(3) Envision Recycling: Baxter provides training to hospital staff to sort plastic intravenous bags, pour bottles, and over pouches. Baxter then collects, transports, and recycles them.

(4) Custom Sterile Packs: Baxter offers customers individualized packs of medical items thereby reducing packaging and saving them disposal costs.

(5) Interwoven: Baxter offers to collect, clean, and deliver disposables such as gowns, drapes and towels so that they can be reused by customers.

(6) Shared-Risk Agreement: The objective is to reduce the amount of products used by Baxter's customers. For every customer, Baxter analyses past patterns of product usage. Then both parties establish a target usage for the future. If the customer uses more than this target, then both Baxter and the customer share the additional costs. If the usage is below target, then both Baxter and the customer share the resultant savings.

Baxter's intra-firm dynamics

Why and how did Baxter decide to market green products? New product launches are typically subjected to rigorous capital budgeting but Baxter did not employ any established procedures to evaluate marketing green products. Power-based explanation is inadequate because marketing of green products has been widely supported within Baxter. There was indeed some initial opposition but policy skeptics did not dig in their heels because they did not view this policy as imposing significant costs on them. Leadership-based theories are most appropriate to explain Baxter's response to marketing of green products. Policy supporters succeeded in convincing their skeptical colleagues that the demand for Baxter's green products is not a passing fad; it reflects fundamental changes in the ways customers prioritize the desired attributes of any product. They pointed out that new laws and regulations (see below) would create incentives for customers to demand products that have less packaging or that do not require polyvinyl chlorides (PVCs) for their packaging. Thus, policy supporters created a consensus about marketing green products, even though they did not employ well-established investment appraisal procedures.

To understand the policy dynamics within Baxter, it is instructive to examine its external environment as well. Many recent environmental laws and regulations have sought to promote the reduction, reusage, and recycling of resources. The objectives of such laws are twofold: to reduce the volume of wastes, particularly solid wastes, that are overwhelming landfills, and to conserve raw materials. One of the strategies has been to enact "take-back" legislation that requires manufacturers or retailers to take-back packaging at no additional cost to consumers. Such laws are intended to create incentives for manufactures to reduce packaging. Europe, in particular, has pioneered such "take-back" legislations. In 1981, Denmark enacted a law requiring manufactures to use reusable beer and soft drink containers and requiring beverage retailers to collect all such containers regardless of whether or not they sold them.[13] In 1991, Germany enacted its own version of "take-back" legislation requiring firms, or their designated agents, to collect and to recycle all packaging. Consequent to a rather successful German program, in 1994 the European Commission issued its own version of "take-back" and recycling directive. In the US as well, pressure has been building for firms to reduce their packaging. As discussed in Chapter 3, in 1991, the Coalition of North Eastern Governors (CONEG) challenged the top two hundred users and producers of packaging in the US to eliminate or to reduce the amount of packaging generated by their companies.

Most of Baxter's green products emphasize packaging reduction which will reduce pressure on their customers' waste streams. This strategy makes business sense because many of Baxter's products require substantial packaging. Note that, in contrast, Lilly markets products that require little packaging. As a result, packaging reduction is almost irrelevant for Eli Lilly.

In response to the CONEG challenge, Baxter committed to a 15 percent reduction in per-unit packaging weight by 1996 with 1990 as the baseline (Baxter 1991b). Baxter's packaging task-force set guidelines on issues such as reducing the use of inks containing heavy metals, modifying Corporate Identity Guidelines for the company logo to allow for the use of recycled paper for office stationary and packaging materials, applying Society of Plastics Industry and American Paper Institute recycling symbols for appropriate packaging, promoting the sale of single package multi-product medical kits, advocating the use of reusable shipping containers, and minimizing the use of chlorine-bleached papers and paperboard in packaging. Thus, policy supporters argued that such initiatives

---

[13] The Danish legislation was criticized however as a non-tariff barrier. Though the "take-back" clause of this regulation still holds, the European Court of Justice struck down the reusable provision (Vogel 1995).

provide Baxter an opportunity to leverage its packaging reduction expertise for marketing green products.

Policy supporters also highlighted the EPA's (then) recent proposal on
stricter air-pollution control standards that would have required an
upgrade in existing medical waste incinerators. Hospitals (Baxter's customers) either have their own incinerators or they send their medical
waste to a third party for incineration. It was estimated that such an
upgrade might cost up to a million dollars per incinerator. Consequently,
if implemented, this policy would shut down about 80 percent of medical
waste incinerators. Policy supporters argued that this proposal would
create incentives for hospitals to demand that Baxter reduce the quantity
as well as certain types of packaging (Baxter 1995b), and such demands
would be in tune with the general concern among health-care providers
about managing medical wastes. As Marshall Abbey, former Senior Vice-
President, General Counsel, and Chairman of Baxter's Environmental
Review Board, noted:

Remember, a large share of Baxter's revenue comes from the sale of disposable
medical products. As soon as these products are used, they become waste. And . . .
waste is becoming a big problem for our customers. So while we are selling products to our customers to satisfy their medical services needs, we are also giving
them waste problems. If we are to keep our customers and attract new ones, it is
critical that we work to minimize these problems. The future of the company
depends on this. (Baxter, 1991a: 3)

In similar vein, while addressing the conference on "Our Environment:
A Healthcare Commitment," held in Arlington, Virginia, March 10,
1992, Vernon Loucks, Baxter's CEO noted that:

[O]ne of the hospital's biggest opportunity involves managing waste. A 500-bed
hospital in the US generates abut 1,500 tons of waste each year – three tons per
bed – and it can spend up to $350,000 getting rid of it. A teaching hospital of that
size typically generates much more waste, and its disposal costs can be well over
$1,000,000. But a hospital can whittle down these disposal costs significantly
through aggressive programs to reduce, separate, and recycle waste.

Baxter's ACCESS program has shown that this approach really works . . . the 288-
bed Children's Memorial Medical Center in Chicago, saved $500,000 in waste
disposal costs over three years. Another customer, Thomas Jefferson University
Hospital in Philadelphia, established a recycling program that produced a
$100,000 savings the first year. (Baxter, 1992b: 4)

Another focus of Baxter's efforts to green its products was reducing the
use of polyvinyl chlorides (PVCs) for packaging. Like many other firms in
the health-care industry, Baxter uses PVCs to package many of its products. PVCs are plastics derived from mineral oil, natural gas, and rock
salt. It is suggested that when incinerated the chlorine in PVCs may

produce dioxins. However, there is no statistically proven relationship between the amount of PVCs incinerated and the levels of dioxin produced. Dioxins, the generic name for a group of seventy-five chemicals known as chlorinated dibenzo-p-dioxins, are shown to cause cancer in some laboratory animals. As a result, there is immense pressure from environmental groups for regulating, if not banning, the use of PVCs. The medical waste incinerators are considered one of the biggest sources of dioxins, primarily because they incinerate PVCs. The controversy over PVCs created incentives for these customers to demand that Baxter should reduce, if not eliminate, the use of PVCs to package its products (Baxter 1996a).

A similar controversy has been plaguing di-2-ethylhexyl phthalate (DEHP), a plasticizer for PVCs. Some scientists claim that substances such as DEHP that exhibit estrogen mimicry may substantially reduce sperm counts and cause other reproductive problems. A widely discussed book, *Our Stolen Future* (Colborn 1996), argues along similar lines. Again, though there is no conclusive evidence on the impact of DEHP on reproductive issues, nevertheless, there is intense pressure by environmental groups to regulate, if not eliminate, their use. Some European countries such as Germany and Denmark have already taken a position against DEHP. This controversy again created incentives for Baxter to reduce its packaging and the use of PVCs (Baxter 1996c).

Though policy supporters highlighted that new laws and regulations create incentives for marketing green products, they also invoked normative factors and pressures from non-governmental actors to generate support for these policies. They emphasized that as a health-care firm Baxter was responsible for safeguarding the health of the environment as well. Baxter's CEO Vernon Loucks, almost acting as their chief spokesman, noted:

So there are good business reasons to invest in the environment. There are ethical reasons, as well. Environmental protection lies at the very core of our social duty. Our industry is dedicated to preserving and improving the health of mankind. . . . Indeed, our industry has a special connection to life that no other industry shares. But we cannot preserve that trust and uphold that bond if we ignore the environmental consequences of our action. *We cannot fight for life at one moment and destroy it the next. We cannot help those at our front door and harm them by the wastes we send out the back.* (Baxter 1992a: 5–6; italics mine)

This book has also suggested an important role for non-governmental external stakeholders in encouraging Baxter to market green products. One way in which such actors exercised influence was by communicating their demands and expectations to Baxter. Actors who supported marketing green products publicized such communications within Baxter, particularly by getting them published in the in-house environmental newsletter *Baxter Environmental Review*. Let me quote from letters written by three different categories of actors: a household using Baxter's product, a hospital, and an investment firm.

Jim and Sarah Tennessen of Menomonie, Wisconsin, sent this letter to Caremark, a division of Baxter:

Dear Caremark:

I am writing you with a challenge. Caremark has a reputation for innovation and quality. I would like to see Caremark apply this innovation towards preserving the quality of the planet we live on. I know of no home care provider who has any type of recycling effort in place at this time. I see this as a tremendous opportunity for Caremark to take on an industry role, not only improving the quality of life for its customers, but maintaining the quality of life for all mankind. Caremark, with its home delivery network, is well-positioned to capitalize on this opportunity.

In a single week, our four year old's HPN [TPN] therapy creates approximately six cubic feet of waste, consisting of seven 2,000 ml I.V. bags, three 250 ml lipid bottles, ten line sheets, seven vitamin and sterile water vials, 21 syringes, 24 needles, 16 face masks, 67 alcohol swabs, four split bandages, two gauze bandages, one role of tape, plus all the miscellaneous packaging for these items . . . . (Baxter 1991a: 1)

Similarly, Mr. David Baker, Director of Materials Management, St. John Mercy Medical Center, St. Louis, wrote:

One of our employees recently asked our president whether he knew the environmental records of the many vendors we deal with. To that end, we would like to request information from your firm about the composition of the products you sell, the functionality of your products and services – anything you do environmentally sound. We would appreciate sharing this information with our employees. (Baxter 1991a: 2)

Scott Fenn, Director of the Investor Responsibility Research Center, a non-profit firm representing 400 investment institutions, including many large investment funds, sent a survey to Baxter seeking information:

Has your company introduced or modified any product or service in the last three years specifically to address environmental concerns? Please indicate the nature of the environmental or financial benefits the company hopes to achieve (Baxter 1992a: 2).

To conclude, Baxter is marketing green products because it could build on its competencies in recycling and packaging reduction, its customers value the benefits of green products for environmental or profit reasons, and customers believe that Baxter's products deliver these benefits. In contrast, Lilly has few business reasons to market green products since most of its products are ethical drugs whose purchases depend on factors such as efficacy and price. These drugs also have small dimensions and hence there is not much scope for reducing their packaging. In addition, these drugs are single-use products that cannot be repaired, reconditioned, remanufactured, or reused.

Since Baxter has not done any investment analysis to justify marketing green products, efficiency-based theories offer little help in understanding Baxter's internal policymaking on this subject. Baxter markets a range of green products and this book has explained the adoption of this policy by employing leadership-based explanations. To persuade the organization to market green products, policy supporters highlighted various benefits of this policy.

First, they emphasized its normative aspects: products of a health-care company should not damage the health of the environment. Second, since many non-governmental actors were asking for information on Baxter's green products and communicating their expectations on this subject, the policy supporters highlighted the goodwill benefits of marketing green products. Third, and most important, policy supporters emphasized that marketing green products will eventually be profitable for Baxter since such products provide quantifiable savings to customers. As discussed earlier, by emphasizing recycling and packaging reduction as key benefits, Baxter's green products help customers in lowering their waste disposal costs. Further, these products also proactively reduce PVC incineration.

These benefits are valued by Baxter's customers in light of the EPA's proposal on upgrading air-control standards for medical waste incinerators. If implemented, this policy would require expensive refitting of such incinerators. Consequently, many of them will shut down and the remaining ones will begin charging much higher prices for incineration. Therefore, the policy supporters emphasized that the demand for green products will be primarily driven by an unmet need: customers' desire to reduce medical waste disposal costs. However, these benefits were not obvious to many managers within the organization. Thus, adoption of policies on marketing green products is best explained by focusing on the role of the policy supporters who explained the business potential of such products, and as a result, generated consensus on adopting such policies.

5   Environmental audits

Both Baxter and Lilly regularly audit their environmental programs. They have internal audits teams that audit these programs at various levels: corporate, division, and facility. Baxter also invites external actors to conduct some of its environmental audits. It invited Arthur D. Little (ADL) to help define state-of-the-art environmental standards and then audit its environmental program against these standards. Further, Baxter even makes such audit findings public.

Environmental audits have become controversial in recent environmental policy discussions. Enforcing environmental laws is a key challenge for state and local government agencies. Often, these agencies are inadequately staffed. Further, employee motivation is low since they are not well-paid and they receive little training for upgrading their skills. In addition, the sheer complexity of most environmental laws and regulations requires that these agencies devote significant proportions of their resources to processing permit applications from the regulatees rather than monitoring and enforcing compliance. Consequently, even in instances of alleged non-compliance, state and local-level regulators are often handicapped by lack of resources to document their cases against the alleged violators.

Environmental regulators have two routes by which to address this issue. First, to relax the resource constraint, they could ask for increases in their agencies' budgets. This would enable them to hire more people, pay them higher wages, and invest in their training. As a result, agencies would potentially become more effective in enforcing environmental laws. This route operates from the supply side: that is, increasing the supply of regulatory services by budgetary increases. This route, though attractive, is difficult in an era of downsized governments, increasing unfunded mandates to the states, and constant criticism of the adversarial relationship between regulators and industry (Chandler 1980; Marcus 1984; Vogel 1986).

The second route to monitor and prosecute non-compliance operates from the demand side: reducing the demand or decelerating the increases in demand for enforcement and monitoring services in the face of declining, constant, or gradually increasing supply. Such demands are being articulated by a variety of actors such as common citizens, environmental groups, media, and legislators. To meet the demand for these services, regulators can propose to create new institutions (such as self-audits) that generate incentives for the regulatees to self-monitor and voluntarily comply with the laws.

The idea behind environmental audits is that if regulatees can create credible management systems to monitor their own compliance and to correct the management systems that led to non-compliance, the demand for enforcement from the regulators will be reduced. After all, a cleaner environment is the end product of enacting complex and stringent environmental laws and setting up regulating agencies. If regulatees can achieve this end product by themselves, then there would be less demand for external enforcers. As a secondary benefit, since many regulatees feel threatened by the idea of regulators knocking at their doors and peering at their compliance records, this plan will also lessen the confrontational relations between industry and government. Thus, this self-monitoring route seems to be win–win for the regulators and the regulatees.

Unfortunately, the situation is complicated by many issues. Suppose a firm conducting a self-audit of its environmental program discovers violations of environmental laws. Should this firm report these violations to regulators? Will these audit findings be self-incriminatory: will the regulators use them as evidence against this firm in civil, criminal, or administrative proceedings? If so, firms have few incentives to conduct self-audits and to report violations to regulators.

Further, suppose this firm does detect violations and does not report them to regulators or make this information available to the public. Can such information be subpoenaed? Again, firms have few incentives to undertake internal audits and to create evidence that could potentially be used against them.

In the early 1990s, some state legislatures led by Oregon, began taking a serious look at passing laws that would create incentives for firms to undertake environmental audits.[14] By the middle of 1995, about twenty states had enacted legislation that encouraged firms to undertake environmental audits. The passing of such legislation was not well-received by the EPA, the Department of Justice, most environmental groups, and some sections of the legal community. The EPA threatened to withdraw the enforcement authority of the delegated federal laws, particularly the Clean Air Act and the Clean Water Act, from states that had passed such legislation (*New York Times* 1996a) and interestingly, some regulatees like Intel Corporation supported the EPA's stand. Needless to say, the EPA was widely criticized. Senator Enzi chided the EPA by noting:

It is particularly galling to hear the EPA proclaim these state environmental laws ineffective. Given the federal government's behavior, how can they possibly be a brilliant success? Determined to extinguish these laws, the EPA and the Department of Justice have stated their intention to proceed with investigations of companies that use environmental self-audits – rendering these state laws moot. The feds have also threatened the states with loss of what little control they have in

---

[14] Most laws stipulated that the attorney–client privilege is inapplicable if: (1) the court determines compelling reasons to disclose such information; (2) the firm has abused this privilege for a fraudulent propose; or (3) this information suggests immediate danger to public health or the environment outside the facility.

the administration of the Clean Air Act and Clean Water Act costing them millions of dollars, if they don't stop their environmental audits. This is blackmail.... *And it is effective blackmail indeed, for until these audits laws are allowed to function as intended, business will never use them to their full promise.* (1997: A 18; italics mine)

The EPA's critics also challenged its opposition to granting attorney–client privilege for self-audits by regulatees. They pointed out that the Federal government grants such privilege for self-audits in areas such as the Equal Credit Opportunity Act and the Federal Aviation Administration. Responding to such criticism, in December 1995, the EPA issued guidelines on self-policing by regulatees. Though it supported self-audits, the EPA reiterated that it would not grant attorney–client privilege to regulatees. However, it promised not to file for punitive damage (gravity component) if the regulatee notifies the appropriate governmental agency within ten days of discovery of the violation, undertakes follow-up action within sixty days of the audit, provided that there were not any repeat violations at the facility during the following three years nor patterns of violation by the company during the following five years. However, the EPA might still pursue the "economic gain component," that is, recover the gains accruing to the regulatee by violating laws. The EPA also said that it would not pursue criminal action if the violation posed no serious harm, if the management did not conceal or condone such violations, and if such violations were not a result of deliberate managerial blindness (Baxter 1996a; *New York Times* 1996a).

As suggested earlier, most environmental groups opposed granting attorney–client privilege for environmental audits. They argued that such privilege has the potential of being misused by firms: firms can escape prosecution simply by acknowledging that they had violated environmental laws. Consequently, firms would have few incentives to invest resources in implementing laws. Raena Honan, Legislative Director of the Grand Canyon Chapter of the Sierra Club described self-audits as the "Bhopal Bill." She argued that:

Any self-reported violation would be secret and immune from civil plus criminal penalties for anything less than a second-degree murder. . . . We are here at the local level watching out for the public's interest because states do not have a record of protecting their citizens. They have a record of granting corporate welfare and pandering to special interests. (*New York Times* 1996b: A26)

In his letter to the EPA's Administrator Carol Browner, Carl Pope, Executive Director of the Sierra Club, suggested that his group may sue the EPA if the agency did not force the states to comply with the federal environmental laws (*New York Times* 1996a). Some sections of the legal community were also very critical of granting attorney–client privilege to regulatees. Edward Heilig, an Assistant District Attorney in the environmental crimes unit in Suffolk county, New York, noted that:

All they have to do is to stamp their documents with the words "audit privilege," and we won't be able to look at them (*New York Times* 1996a: A16a).

As a result, Heilig argued that his office would be denied access to information required to investigate environmental crimes. The National Association of Attorneys General identified some legal pitfalls as well (Morandi and Pascal 1995). They criticized the fact that the onus of proof had been shifted to the regulators who were now required to demonstrate that a violation had occurred before receiving access to audit reports. Further, the judges and not the juries (juries tend to be less sympathetic towards firms) would determine whether a firm abused its audit privilege. As a result of these hurdles, regulators would have to devote more resources towards legal battles for having access to the necessary documents on environmental violations. Consequently, the major objective of the new audit laws – reducing enforcement responsibilities of the regulators – would not be served.

To summarize, though some states have enacted legislation granting attorney–client privilege to environmental audits, the EPA, many environmental groups, and some sections of the legal community vociferously oppose such provisions. The EPA does not recognize state laws that grant such privileges. As a result, even if states grant these privileges, the EPA can still have access to the audit information and employ it as evidence against the regulatees for violations under federal laws. Thus, firms have few incentives to establish environmental auditing systems, to undertake audits, and to report violations that have been detected by such audits. Further, if they conduct audits, we expect them to focus on management system audits and not on compliance audits (discussed below); that is, they will focus on assessing whether appropriate environmental management systems are in place, and not on whether any law has been violated. Even if they document that some management systems are not in place, this will not necessarily create self-incriminatory evidence.

Baxter and Lilly: policy dynamics

Baxter and Lilly are committed to conducting environmental audits and have established active programs to audit their facilities, divisions, and corporate groups. Eli Lilly's *Environmental Policy and Guidelines* considers auditing an integral part of Lilly's environmental policy:

Table 4.12. *Internal audits performed at Lilly*

	Lilly facilities	Non-Lilly facilities
1992	8	15
1993	8	23
1994	12	12
1995	15	5

Source: Lilly (1997b).

The company will regularly assess and report to management and the Board of Directors on the status of its compliance with this policy and with environmental laws and regulations. (Lilly 1996b).

Likewise, Baxter's *Environmental Policy* states that:

Corporate environmental personnel, division, and facilities will provide coordinated, effective environmental training, awareness and audit programs wherever relevant (Baxter 1993/94).

Both firms conduct two types of audits: compliance and management systems. Let me give an example to differentiate these audits. Assume that, while auditing a facility, the auditors discover an unmarked drum containing hazardous wastes. If this were a compliance audit, the auditors would inquire whether there are any law(s) that require drums containing hazardous substances to be appropriately marked. If so, which laws, and who is to be held responsible for this violation? In contrast, if this were a management systems audit, the auditors would inquire whether the facility has management systems to ensure proper labeling of drums containing hazardous wastes. If so, why did they fail, and how can the system design be modified to prevent reoccurrence of such failure?

Lilly's environmental auditing program was formally initiated in 1989. In that year, Ralph Hall of the Corporate Environmental Legal group led Lilly's first internal audit that focused on the compliance of the Tippecanoe facility with the requirements of the Resource Conservation and Recovery Act (RCRA). During 1989–1994, Lilly focused primarily on compliance audits. Since 1994, systems audits have also gained prominence. As shown in table 4.12, since 1992, Lilly has also been auditing its suppliers, waste-treatment sites, and third-party manufacturers. In 1996, Lilly also began auditing its non-US facilities. As of 1997, Lilly relies on internal audits only. Lilly's audit teams are comprised of managers from its corporate environmental groups (Environmental Affairs and Environmental Legal) and environmental managers from plant sites outside of the sites being audited.

As discussed in chapter 3, in the late 1980s, Baxter began revamping its environmental program. Baxter initiated its environmental audit program in 1988. It hired Arthur D. Little (ADL) to assist the Environmental Legal Affairs group in developing an audit protocol. ADL is well known for its expertise in environmental auditing; it has been doing such audits since 1977 and conducts about 400 environmental audits annually. Once the protocol was developed, Baxter tested it by conducting pilot audits in three facilities: Bentley facility in Irvine, California, Pharmaseal facility in Jacksonville, Texas, and the Dade-East plant in Miami, Florida. Overall, the protocol was found to be excellent, requiring only minor changes (Baxter 1988). As a result, the environmental protocol was adopted firm-wide.

In 1990, Baxter took another critical step towards strengthening its environmental auditing program. As discussed in chapter 3, Baxter's senior management decided that Baxter would have a state-of-the-art environmental program. It invited ADL to help them define state-of-the-art environmental standards for firms in similar risk categories, and then to audit Baxter's programs against these standards. If a facility met these standards, it was certified as having achieved state-of-the-art standards. ADL also noted that since state-of-the-art standards are dynamic, periodic assessments are required to recalibrate them, and then to evaluate whether facilities meet the new standards. Consequently, ADL audited again in 1993 and 1995. In 1995, it co-audited with Skadden Arps.

Baxter also reported a summary of audit findings in its annual *Environmental Performance Report*. For example, it reported that in 1995 there were 464 audit findings. Of these, fourty seven were considered major items defined as serious deficiencies that pose significant risks to the company: failure to comply with an important regulatory requirement, or lack of a key environmental management system (Baxter 1996a: 18).

Analysis

Power-based and leadership-based theories together explain: (1) why Baxter and Lilly undertake environmental audits, and (2) why Baxter does, and Lilly does not, invite external auditors to undertake such audits. Baxter's adoption of environmental audits, especially its invitation to ADL to help define state-of-the-art standards, and the accelerated schedule of its implementation, was met with internal opposition. The Divisional and facility managers were uncomfortable in committing to ambitious targets, and to inviting external auditors. Given this persistent opposition, and at the same time, a commitment to implement these policies in a short time schedule, policy supporters resorted to having these policies imposed on the organization. Baxter's response to environmental audits is therefore best explained by power-based explanations.

Lilly's adoption of environmental auditing policies is best explained by leadership-based theories. Policy supporters managed to create a broad degree of organizational consensus on a new way of evaluating performance on environmental issues. Their strategy was to convince skeptics, whether at facilities or in the corporate groups, of the long-term benefits of having strong internal environmental audit programs, notwithstanding the EPA's opposition. Importantly, skeptics did not view environmental audits as inflicting losses on them. Consequently, they did not have incentives to dig in, and resolutely oppose these policies.

Lilly's environmental audit program has survived and prospered for many reasons. First, this program has had very committed supporters, particularly Joan Heinz of the Corporate Environmental Legal group, and John Wilkins and Jonathan Babcock, both of the Corporate Environmental Affairs group. These individuals ensured that environmental audits are employed not merely as tools of corporate oversight, but also as tools of system improvement. As one of the managers observed, the audit group is careful not to appear to the facilities as if they are the internal EPA. As a result, facility managers have not felt threatened by the prospect of having their facility-level environmental programs audited.

Second, these managers convinced the organization that Lilly's environmental management systems were fairly robust and that they do not expect audits to uncover any major violation that the EPA or other regulators would not have discovered in the normal course of investigation. As discussed earlier, since the EPA does not recognize attorney–client privilege for environmental audits, firms are wary of conducting audits and creating self-incriminatory evidence.

Third, these managers emphasized that, even though audits were opposed by the EPA and environmental groups, in the long run these actors would begin to appreciate their usefulness. Thus, these individuals succeeded in convincing skeptics within Lilly that a proactive policy to establish an internal auditing program may have some short-term risks but significant long-term payoffs.

The critical watershed was the first audit of the Tippecanoe facility conducted in 1989. Since it revealed certain discrepancies and shortcomings (primarily caused by inadequate staffing), it provided impetus to supporters of environmental programs within Lilly to ask the Environmental Management Committee (EMC) for more resources to strengthen audit programs. Robert Williams, the Vice-President of

Environmental Health and Safety (EHS) and Quality Control/Quality Assurance (QC/QA), and Daniel Carmichael, Deputy General Counsel, were key actors in convincing the EMC to institutionalize environmental auditing. In the highest positivist tradition, they argued that Lilly should be able to measure its compliance with its *Environmental Policy and Guidelines* and that environmental audits were useful tools for this task. Thus, policy supporters were able to leverage the outcome of the first audit and convince the policy skeptics of the benefits of institutionalizing this program.

Fourth, the environmental audit group has constantly explored ways to make environmental audits less burdensome for Lilly's facilities. Environmental auditing is a cumbersome and documentation-intensive process primarily because of the complexity of environmental laws. As a result, facility managers spend a significant amount of their time preparing for audits and implementing the findings of the auditors. Facilities are subjected to other audits as well, such as financial, QC/QA, and health and safety. Consequently, one of the major concerns of policy skeptics was that environmental audits would create unnecessary diversions for the facility managers, thereby impeding them from focusing on meeting production targets. In response to these concerns, the environmental audit group has been very sensitive to the needs of the facilities when scheduling their audits. To minimize disruption in the facility's routine operations, they communicate their audit check lists in advance. Currently, they are exploring with the QC/QA group whether joint environmental/QC/QA audits can be conducted. All such actions signal to the organization the sincerity and constructive intent of environmental auditors. As a result, the policy supporters generated a broad degree of organizational consensus on the usefulness of environmental audits, notwithstanding the hostile stance of the EPA towards it.

Unlike Lilly, why did Baxter's policy supporters rely on power-based processes? Inviting external auditors was clearly a very controversial issue within Baxter. Policy supporters argued that inviting ADL was a demonstration of Baxter's commitment to making its environmental record accessible to external actors and as a testimony of confidence in its environmental programs. They further suggested that since ADL is the oldest, and probably the most established environmental consulting firm, its input in defining state-of-the-art standards will be most helpful. Consequently, external actors will view the state-of-the-art standards defined by ADL as credible and not self-serving. ADL's audit will also provide a useful external perspective on how Baxter could improve its environmental performance.

However, the proposal to invite ADL was internally opposed on four counts. First, the skeptics were concerned about the fuzziness of attorney–client privilege for environmental audits. Second, some managers were uncomfortable with the notion of state-of-the-art standards and committing their facilities to these standards on an accelerated implementation schedule. Some of them questioned the very possibility of defining state-of-the-art standards. Third, some facility managers viewed internal audits as yet another coercive tool employed by the corporate office to make them fall in line. Like many other large firms, Baxter has a tradition of contestation between the corporate office and the facilities. Environmental audits were yet another arena of such contestations. Fourth, since the cost of hiring external auditors was to be eventually borne by facilities, facility managers questioned the need for inviting external auditors at all.

The policy supporters realized that it would be difficult to generate quick consensus within Baxter on environmental audits. I believe a key reason was that the pace and extent of change in environmental programs required to implement state-of-the-art standards were significant. In the mid 1980s, Baxter had a faltering environmental program, and in the 1990s, it decided to have the best program in the industry. Instead of slow and consensual change, policy supporters wanted rapid change. Since such change was resisted, it had to be mandated from the top management. In October 1990, Vernon Loucks, Baxter's CEO, in his address to Baxter's annual internal environmental conference, observed that:

Our overall goal as stated in this policy is this: to develop an environmental program that will be considered state-of-the-art among the Fortune 500 companies. By state-of-the-art we mean among the best in terms of day-today problem solving and compliance, and in managing our short- and long-term environmental risks. We also mean among the best at minimizing the generation and discharge of waste and other adverse impacts on the environment. Our focus is not just on our operations, but on our products and services as well. (Baxter 1990b: 3)

Loucks went on to emphasize that dissent would not be tolerated. He said:

If anyone doubts we are serious about our new policy and goals, let me put that to rest now. We will monitor the company's environmental program, and the program of each division thorough and annual state-of-the-program report. I have requested that this report be presented to the Senior Management Committee and our board of directors. *Let me assure you, inadequate progress will not be tolerated.* (Baxter 1990b 3–4; italics mine).

To reinforce Vernon Louck's message, during the 1991 annual environmental conference, Marshall Abbey, the Senior President, General Counsel, and Chairperson of the Environmental Review Board observed:

Why go to all this trouble when we could probably get by with much less? The answer to this question lies in another question: what kind of company do we want to work for?

Do we want a company that is concerned with people's health and their right to live in a healthy environment? Do we want a company that stands apart from pol-luters and violators of environmental law; a company that is doing something about the environment, not just giving a lip service? *Vern Loucks wants this kind of company. The Environmental Review Board does too.* And if that's the kind of company you want, I think you do, then you must find one like Baxter that is pressing for state-of-the-art in the environmental area. That's what our goal is all about. *And that's my answer to the skeptics.* (Baxter 1992b: 4; italics mine).

Highly visible support from the CEO and the Chairman of the ERB clearly signaled that Baxter's senior management attached high priority to the environmental audit program. As a result, Type 2 policies on envi-ronmental audits, both internal and external, were adopted inspite of opposition from policy skeptics. Again, the learning from this case is that if policy supporters are convinced of the usefulness of a given program, and have the organizational clout to impose their preferences, the program would be implemented.

To conclude, although there was legal controversy on the status of attorney–client privileges for environmental audits, Lilly and Baxter established strong audit programs. In Lilly, policy supporters succeeded in inducing consensus because the pace of change was gradual and audit standards were defined internally. Importantly, their firm was committed to implementing all applicable environmental laws, had established systems, and had invested resources for achieving them. Consequently, skeptics bought into the argument that environmental audits were not expected to uncover major violations that could create problems for them.

In Baxter, the environmental audit program was adopted through a power-based route. Some managers were skeptical about the pace of implementing the state-of-the-art standards, and that these standards were being defined with the help of ADL. Since policy supporters did not succeed in generating a quick consensus on these issues, they resorted to having the policy imposed by the top management.

6   ISO 14000 environmental management standards[15]

So far the book has examined cases where Baxter and Lilly adopted Type 2 policies due to leadership-based and/or power-based processes.

---

[15] This section draws on Prakash (1999a).

This section examines a policy where neither Baxter nor Lilly adopted a Type 2 project – the International Organization for Standardization's (ISO) ISO 14000 environmental management standards. For the time period of this book's study (mid 1975–1996), both Baxter and Lilly adopted a wait-and-see policy towards ISO 14000. This is puzzling because: (1) both firms had ISO 9001 quality certification and therefore had experience with implementing other ISO standards; (2) they had well-established environmental management systems; and (3) economic globalization, and expanding outside the US was important to their respective corporate agendas and ISO 14000 could remove a potential hurdle to tap foreign, especially European, markets. Baxter's resistance to ISO 14000 is especially puzzling since it was (and still is) committed to having a state-of-the-art environmental program.

This book attributes the non-adoption of this policy to the absence of either power-based or policy-based processes to work in these firms. Policy supporters could not capture the top management on this policy issue, it did not have the hierarchical power to impose these policies on the skeptics, and it could not muster the ability to induce consensus over these policies. This was due to factors both external and internal to firms. Regarding the former, less hostile external regulatory and legislative environments for the firms in the mid 1990s and the lukewarm support of the EPA to ISO 14000 reduced the incentives for policy skeptics to reassess their opposition or the top management to impose this policy. Internal factors include the emergence of a "green-wall" (see below) in these firms that made their managers less enthusiastic about Type 2 policies, and the firms'(then) extant investments in well-functioning environmental management systems that reduced the incremental benefits of adopting ISO 14000.

Founded in 1946, the ISO is a Geneva-based non-governmental organization that promotes the development and implementation of voluntary international standards. Since the mid 1990s, this organization has been developing a series of environmental management systems standards – ISO 14000 – for manufacturing as well as service organizations. ISO 14000 attempts to replicate the success of ISo 9000 quality control and quality assurance standards in environmental management that were introduced in 1987.[16] ISO 14000 neither replaces industrial codes of practices such as the CMA's Responsible Care nor local laws and regulations. In fact, the latter are treated as minimum requirements

---

[16] The ideas embedded in ISO 9000 are not new or radical. They share commonalities with Deming's fourteen points (1982, 1992), Total Quality Management, and Malcolm Baldrige National Quality Award criteria.

Table 4.13. *ISO 14000 series: an overview*

ISO series	Description
ISO 14001	Environmental Management Systems: Specifications with Guidance for their Application
ISO 14004	Environmental Management Systems: General Guidelines on Principles, Systems, and Supporting Techniques
ISO 14010	General Principles of Environmental Auditing
ISO 14011	Audit Procedures
ISO 14012	Qualifications Criteria for Environmental Auditors
ISO 14024	Environmental Labeling
ISO 14031	Environmental Performance Evaluation
ISO 14040	Guidelines on Life Cycle Assessment
ISO 14050	Terms and Definitions

Source: Adapted from Puri (1996: 18).

for designing ISO 14000 systems. As shown in Table 4.13, the ISO 14000 series consists of one mandatory standard (ISO 14001) and several non-mandatory standards. ISO 14001 being a compliance standard needs certification by an external auditor: any facility seeking ISO 14000 certification is required to demonstrate that its environmental management systems meet the criteria specified in ISO 14001. In contrast, other ISO 14000 standards are only guidelines and do not require certification.

Baxter and Lilly: response

During the period of this study (mid 1975–mid 1996) both Baxter and Lilly decided against becoming early adopters of ISO 14000. They adopted a wait-and-see attitude and neither of them mandated that their facilities should adopt ISO 14000. This is not to say that they rejected the possibility that their facilities would eventually be ISO 14000 certified. In fact, these firms became "ISO ready" so that, if required, they could have the ISO certification at short notice. Towards this end, they undertook a "gap-analysis" by comparing their environmental systems with the ISO 14000 requirements and identifying areas where their systems needed modification (appendix 4.4). Further (beyond the period of this study), some of Baxter's facilities received ISO 14000 certification. As of July 29, 1997, seven of its 120 facilities are ISO 14000 certified. Lilly's Ireland facility received the ISO 14000 certification in late 1997. Baxter has also commissioned pilots on ISO 14000; five of its facilities will receive the

ISO certification and Baxter will use their experience to evaluate future courses of action (Baxter 1995a). Nevertheless, neither of these firms mandated that all of their facilities should have the ISO certification within a given period of time. This is an interesting puzzle given that these firms have been in the forefront to adopt beyond-compliance policies and have committed millions of dollars to Type 2 policies.

Policy dynamics and analysis

ISO 14000 supporters in Lilly and Baxter could not (and cannot) justify it on the basis of standard investment appraisal procedures such as capital budgeting. They did not have the power to mandate its adoption or the persuasive abilities to convince policy skeptics about its merits. The question then is: unlike other cases examined in this book, why did the power and/or leadership-based processes not succeed for ISO 14000?

To understand Baxter's and Lilly's responses, it is instructive to examine the costs and benefits to individual firms of adopting ISO 14000. Firms face three kinds of costs. First, they bear the costs of creating, adopting, and implementing new management systems. Since ISO 14000 requires extensive documentation many firms may not have the resources for meeting its requirements. Second, firms bear the costs of third-party certification. These costs are significant: typically around $20,000 per facility per audit (Baxter 1996a). Thus, for a firm such as Baxter that had over 200 facilities, third-party certification may cost up to $4 million.[17] Third, firms that invest significantly in research and development are apprehensive about giving outsiders (that is, the third-party certifiers or auditors) access to confidential information. Consequently, to guard against industrial espionage, such firms will need to expend resources for establishing internal protocols on information disclosure.

Firms also face uncertainty on the responsibilities of third-party certifiers. It is not clear if the findings of such certifiers are to be made public and whether the certifiers are required to report instances of environmental violations to the regulators. If the latter is true, it creates disincentives for managers to share compliance information with certifiers. This is important because under many laws, civil and criminal charges can be brought against individual managers for non-compliance with environmental laws. Further, as discussed in the section on environmental audits, the audit findings may not be protected by the attorney–client

---

[17] Another study estimates that for a firm with twelve facilities, overhead and certification expenses could amount to $1 million per annum; certification expenses alone to be around $25,000 per facility (NSF International 1996).

privilege. This again creates disincentives for US-based firms to invite external certifiers to audit their environmental programs.

ISO 14000 standards also create multiple benefits for firms. ISO 14000 is a key policy response by business firms to environmental issues (Cascio 1994). Firms as a group receive two categories of benefits. First, ISO 14000 replaces multiple environmental standards with a single standard. By increasing transaction costs, multiple standards act as non-trade barriers to international trade. In the 1990s many national and regional environmental standards such as the UK's British Standard 7750, the Canadian Standards Association's CS AZ750–94, and the European Community's Eco-label and Eco-Management and Audit Standards have been proposed. Since these standards have not been fully harmonized, they impede trade. By superseding these standards, ISO 14000 will reduce trade barriers.

ISO 14000 could preempt industry-unfriendly standards made by bureaucrats, thereby reducing the scope of industry-hostile monitoring and enforcement by the regulators and environmental groups. The adversarial relationship between business and government in the United States, especially on environmental issues, is well documented. Though this "adversary economy" co-exists with numerous instances of "capture" (Stigler 1971) of governmental instrumentalities by business firms (Kolko 1963; Bernstein 1955), nevertheless firms generally view government-sponsored standards as being inefficient and "industry unfriendly" (Cascio 1994; International Organization for Standardization 1995a, 1995b). This is because such standards are written by bureaucrats seeking to demonstrate to their constituencies (politicians as well as environmental groups) that they are being "tough with polluters." Many managers believe that firms could deliver comparable levels of environmental performance at a lower cost if they were themselves to write such standards. ISO 14000 provides opportunities for firms to participate in the writing of efficient and easy-to-implement environmental standards.

The ISO 14000 certification process potentially provides credible third-party certification of a firm's environmental management systems. Many managers perceive that certifiers often have a problem-solving attitude. Contrast this with the managerial perceptions of the attitudes of state and federal environmental regulators. For example, inspectors of environmental regulatory agencies are prohibited to give constructive feedback or written comments to firms on the positive aspects of their environmental programs. This has been repeatedly pointed out during my interaction with many firms, including Baxter and Lilly. Similarly, ISO 14000 may reduce public demand for new and more stringent environmental regulations. A widespread acceptance of ISO 14000 may mollify demands of some environmentalists for uniform process standards, which are currently disallowed by the World Trade Organization (formerly, the General Agreement on Tariffs and Trade).

Although significant benefits may accrue to firms as a group, individual firms may have incentives to free-ride and not to invest in adopting ISO 14000 standards. As discussed in chapter 2, an important issue then is whether the institutional design of ISO 14000 creates sufficient net *excludable* benefits for firms to adopt ISO 14000. Further, are these benefits quantifiable, enabling this project to meet the criteria of formal project assessment procedures? The excludable benefits are the following.

First, firms with ISO 14000 certification will adhere to and register for only one standard and not multiple national standards. This will reduce the transactions costs of multiple certifications and enable firms to tap economies of scale. Second, as suggested earlier, the European community may require firms to get ISO 14000 certification to qualify for governmental purchases (Johnson 1994). Hence ISO 14000 certification provides excludable and quantifiable benefits for firms with a business presence in Europe. Note that a similar debate took place in the United States on ISO 9000. Since the European Community had adopted ISO 9000 (and called it EN 29000), it was feared that this may serve as a non-trade barrier for US exporters. Therefore, US firms wanting access to European markets have incentives to get the ISO 9000 certification. For firms catering predominantly to non-European markets, however, ISO 14000 will bring little excludable and quantifiable benefits. Third, firms may have opportunities to reduce costs by improving environmental performance. Firms may lower their insurance costs and have easier and cheaper access to credit and loans by documenting sound environmental practices (Schmidheiny and Zorraquin 1996). Fourth, having ISO 14000 certification demonstrates the intent of a firm to follow environmental laws and regulations in a systematic manner. Even though the EPA does not grant attorney–client privileges, other environmental regulatory agencies may take a less harsh view of minor violations by firms that have such well-established environmental management systems. Fifth, since ISO 14000 may serve as a blueprint for future environmental laws and regulations, its certification may prepare firms for implementing future laws. Finally, ISO 14000 may equip firms for participating in law-making processes. Often, regulatory agencies invite firms, citizen groups, and industry associations to comment on the proposed drafts of a new environmental law or regulation. For example, an internal document of Lilly reported that:

Further gains have been made in our efforts for a proactive involvement in legislative and regulatory issues at the federal and state level . . . Lilly has been particularly active in . . . the EPA's Clean Air Act Advisory Committee. We played a major role during 1992 in developing the proposed regulations for implementing Title V (Operating Permits) under the 1990 Clean Air Act.

At the state level, we have been a major player in the most significant emerging environmental legislations. The principal focus has been on enabling legislations for state implementation of the Clean Air Act. Other activities have centered on the subjects such as pollution prevention, the state's proposal for "good character" requirements for permit pursuance, rulemaking process, and adequate staffing of the state's environmental agency. Our regulatory focus has been on air rules dealing with interim construction permits and the definition of construction and modification. We continue to assist in the development of the "Voluntary Remediation Program." (1992b: 28)

Since the inputs of firms with prior experience with environmental management systems will be more credible, such firms may have opportunities to leverage their expert knowledge acquired during ISO 14000 certification to shape laws and regulations to their advantage.

Then why did Lilly and Baxter adopt a wait-and-see attitude on ISO 14000? Why did power-based dynamics not work? Or, why did some organizational entrepreneurs not take a lead in convincing their skeptical colleagues that having ISO 14000 certification has long-term net benefits? I attribute the absence of power-based and leadership-based dynamics to factors internal and external to firms: the presence of "green-walls" (explained below), extant well-functioning environmental management systems, attorney–client issues, and declining external pressures to "go-green."

There is an emerging literature that suggests that since the mid 1990s "green-walls" have emerged in firms that discourage managers from initiating Type 2 policies. Green-walls manifest in budget and headcount reductions of environmental departments and a reduced representation of senior managers on environmental committees. An article published in *Perspective*, Arthur D. Little's influential management journal, notes that:

A close inspection of today's management of corporate environmental matters indicates that everything is not going smoothly. Although some companies are moving ahead with reliable strategic environmental initiatives, still others are backing away from the broad programs they adopted over the last decade. Why? Because they have hit the Green Wall – they have reached that point at which management refuses to move forward with its strategic environmental program. . . . Early symptoms include . . . deferred decisions because of reduced management support, and an inability to demonstrate return on investment in environmental programs. (1995: 2)

Why has the so-called green-wall emerged only in the mid 1990s and not earlier? How does it impact the adoption or non-adoption of ISO 14000? The green-wall has emerged due to the following factors. First, environmental programs have tended to become victims of their own success. As described in chapter 3, in the late 1970s and the 1980s, firms such as Baxter and Lilly invested significant resources in beyond-compliance environmental programs due to the perceived hostility of the regulators as well as pressures from stakeholders. Cheerleaders, both within and outside firms, proclaiming "win–win" scenarios over-promised benefits of such investments (Walley and Whitehead 1994). Consequently, many environmental programs did not generate expected levels of benefits causing disappointment among many managers. This was also a time when managers were under tremendous pressure to cut costs (see below). Not surprisingly, in the 1990s many managers began closely scrutinizing environmental projects for their impact on quantifiable profits. Supporters of Type 2 policies are, therefore, finding it difficult to convince their skeptical colleagues of the future benefits of such programs.

Second, some environmental managers have adopted cultures that are out of sync with their firm's culture. As one of Baxter's managers put it, "we did not talk the language of business in terms of profits and dollars." In the 1980s, riding on a green-wave, environmental managers felt little need to communicate with managers in other functional areas. Thus the green-wall in the 1990s represents, in part, a backlash from non-environmental managers.

Factors external to firms have also contributed to the emergence of a green-wall. As evident in the initiatives of 104th and 105th US Congresses, there is a political backlash to the perceived excesses of some environmental regulations. The 104th and 105th US Congresses sought to dilute many environmental laws and even cut the budget of the EPA. Though they did not succeed in many of such quests, firms are nevertheless feeling less pressured to gain external credibility by investing in Type 2 environmental programs.

The increased activity in the merger and acquisition markets in the 1990s, from $400 billion in 1990 to about $1.6 trillion in 1997 (UNCTAD 1998), and the perceived pressures of the emerging global economy have made firms very cost conscious. An increased recourse to corporate downsizing for boosting corporate profits has often resulted in the axing of softer areas such as environmental management. This issue has been repeatedly emphasized by managers in Lilly and Baxter. Thus, environmental mangers are increasingly being challenged to make a business case based on quantifiable profits for their beyond-compliance programs. This is not to say that the green-wall encourages firms to violate environmental laws; it only impedes the adoption of Type 2 policies.

ISO 14000 has been critically hurt both by the emerging green-wall and external factors. First, it was initiated in an era when environmental programs are under attack both within and outside firms. Consequently, firms are feeling less pressured to adopt Type 2 policies such as ISO 14000. As I will discuss in chapter 5, external threats are often critical for internal coalitions to develop for championing Type 2 beyond-compliance policies. Skeptical managers are not dismissing ISO 14000 altogether; they are recommending a wait-and-watch policy. The issue is not whether or not ISO 14000 is beneficial in the long run; there is a fair degree of agreement that it will have benefits in the long run. Rather, the debate is when a firm should invest in ISO 14000 certification. ISO 14000 supporters are finding it difficult to convince skeptics that it is worthwhile to be an early adopter.

Second, ISO 14000 offers little value to firms that already have well-established environmental management systems. Such path-dependencies that are emphasized by new institutionalists – past policies influence current and future outcomes – have reduced the incremental benefits of ISO 14000. Many firms already subscribe to some well-recognized industrial codes of practice such as Responsible Care. This again reduces their need to gain external legitimacy by joining another industrial code of practice such as ISO 14000.

This discussion summarizes the mood within Lilly and Baxter on ISO 14000. Lilly's resistance to ISO 14000 can be traced to the factors discussed above. First, Lilly already had a well-established environmental control process: Environmental Quality System (EQS). This system is modeled on the lines of stringent product quality systems employed by pharmaceutical companies for meeting the US Food and Drug Administration's strict guidelines. At the heart of EQS is a computer-based Plant Site Environmental Compliance Listing (PECL) system (appendix 4.2). Lilly implemented this system in 1994. PECL simplifies environmental permits and regulations into an easy-to-use plant site compliance list. As a result, every tank, every pump, and every compliance point is identified for its environmental requisites (Lilly 1995a).

Second, as discussed previously, Lilly is actively implementing a widely accepted code of industrial practice: the CMA's Responsible Care program. Lilly has very active organizational involvement in the CMA with John Wilkins of Corporate Environmental Affairs chairing the Indiana-Ohio-Kentucky chapter of Responsible Care. Although Responsible Care and ISO 14000 have many commonalities, they also

Table 4.14. *Lilly's environmental program staff*

Year	Full-time equivalent in person years	Person years/net sales in $billions
1990	218	52.2
1991	271	59.8
1992	313	63.1
1993	340	65.4
1994	338	59.2
1995	334	49.4

Sources: (1) Lilly (1994a: 31).
          (2) Lilly (1997a).
          (3) Lilly (1992b:10).
          (4) Lilly (1995c:34).

differ on some issues. Unlike ISO 14000, Responsible Care requires that firms undertake community-outreach programs. On the other hand, ISO 14000 requires external certification where as Responsible Care does not. Thus, apart from the costs of getting the ISO 14000 certification, there were additional costs involved for Lilly to modify its management systems and adopt ISO 14000 compatible systems. However, Eli Lilly would have received little incremental external validation for its environmental program by having its facilities ISO 14000 certified.

Third, some sort of a green-wall had developed in Eli Lilly. As shown in table 4.14, since 1993, Lilly's human resource commitment to environmental programs as a proportion of its net sales has declined. Further, as discussed in chapter 3, Eli Lilly's corporate environmental program, which was previously headed by a manager of the rank of Vice-President, is now headed by a manager of the rank of a Director. This again reflects the diminished clout of environmental affairs within the organization. As discussed previously, the emergence of a green-wall is a symptom of a more close-fisted approach to Type 2 policies such as ISO 14000.

Fourth, Eli Lilly has had little experience with external audits of its environmental program. Since ISO 14000 requires third-party certification, some managers were apprehensive about the potential misuse of audit information, especially since the EPA does not grant attorney–client privilege on environmental audits.

Baxter also adopted a wait-and-see policy on ISO 14000. It did not mandate its facilities to seek the ISO 14000 certification. Verie Sandborg (1996), Baxter's Environmental Affairs Manager, emphasized that:

We are going to take a harder look at the requirements contained in the standard and what value ISO 14001 can add to our program . . . We essentially have all the basic components. On the whole, our program is comparable [to ISO 14001]. (1996: 23)

Like Lilly, Baxter has well-established environmental management systems. As discussed in the cases on environmental audits, Baxter regularly invites Arthur D. Little to reassess its state-of-the-art standards and audit its environmental systems against these standards. Therefore, ISO 14000 contributed little to improving the design or working of Baxter's environmental management systems. Further, Baxter's environmental programs had received wide recognition. Consequently, Baxter gained little incremental external credibility by investing in ISO 14000 certification.

A green-wall also impacted Baxter's environmental programs in a subtle way. Throughout the 1990s Baxter had a Vice-President, William Blackburn, heading its corporate environmental affairs. Thus, there was no palpable downgrading of intra-organization clout of the environmental department. Environmental managers also seemed to have preempted a major demand associated with an emerging green-wall: a demand that beyond-compliance projects be justified on business grounds. As discussed in chapter 3, Baxter's Corporate Environmental Affairs calculates the net monetary impact of environmental programs on Baxter's profits. However, this proactive stand on calculating the business impact of policies backfired in the case of ISO 14000; its policy supporters could not make a business case for early adoption. In addition, since there is educed pressure on firms from the external environment for visible environmental initiatives, ISO 14000 policy supporters could not forge a coalition in its support.

To conclude, power-based or leadership-based processes did not work for three reasons. First, these firms had (and have) well-functioning environmental management systems that had wide-spread external credibility. This path dependency made ISO 14000 less attractive to them.[18] Second, neither of these firms faced any external threats that could encourage power dynamics or facilitate emergence of coalitions to push for ISO 14000. Third, both firms faced the emergence of some sort of green-wall. This challenged the adoption of environmental policies, such as ISO 14000, that cannot be justified on the basis of established procedures of investment analysis. Both power-based and leadership-based processes require individuals to take ownership of a given program but the green-wall discouraged managers from championing ISO 14000.

---

[18] There is a well-established literature on the impact of path dependency on adoption of standards. Important works include Katz and Shapiro (1983, 1985), Bessen and Saloner (1988), and Saloner and Shepard (1991).

Appendix 4.1   Six codes of Responsible Care

Code 1. Community awareness and emergency response

Approved: November 6, 1989
Objective: Ensure emergency preparedness and foster community-right-
to-know.
Requirements:
• Develop community outreach programs for communicating information on environment, health, and safety aspects of facility's operations.
• Develop programs for responding to emergencies. Test these programs annually involving all relevant stakeholders.

Code 2. Pollution-prevention

Approved: April 6, 1990 (waste and release practices); September 5, 1991
(waste management practices)
Objective: Promote pollution-prevention and waste-minimization.
Requirements:
• Document waste generation, estimate their releases to various media,
and evaluate their potential health, safety, and environmental impacts.
• Seek employee and public input for developing and implementing
waste-minimization and pollution-prevention policies.
• Emphasize source-reduction and include pollution-prevention as an
objective at the research and development stage.

Code 3. Process safety

Approved: September 10, 1990
Objective: Prevent industrial accidents.
Requirements:
• Develop process safety programs. Document and measure safety per-
formance.
• Audit safety systems. Conduct safety reviews of new-modified facili-
ties before commissioning them.
• Train employees in safety procedures.

Code 4. Distribution

Approved: November 5, 1990
Objective: Minimize risks posed by transportation and storage of chemi-

cals to carriers, customers, contractors, employers, and the environment.

Requirements:

• Evaluate risks associated with existing modes of transportation and distribution.

• Train employees, carriers, and contractors on the regulations and best practices.

• Regularly review the performance and practices of carriers.

• Develop an emergency response plan for dealing with transportation accidents.

Code 5. Employee health and safety

Approved: January 14, 1992

Objective: Protect and promote health and safety of employees and visitors at facilities.

Requirements:

• Develop and review occupational safety systems; audit them; train employees.

• Select vendors and contractors that follow the above guidelines; audit them.

• Investigate trends in workplace illness, injuries, and accidents.

Code 6. Product stewardship

Approved: April 16, 1992

Objective: Promote safe handling of chemicals from their initial manufacture to their distribution, sale, and disposal.

Requirements:

• Develop a corporate plan on product stewardship.

• Incorporate environmental, health, and safety concerns on the product and process development stage.

Appendix 4.2 Baxter's environmental awards

1988–1993

BAXTER OPERATIONS received 37 environmental awards.

1994

BAXTER
The Issue Exchange
W. Howard Chase Award for Excellence in Issue Management

CAROLINA, PUERTO RICO
Environmental Quality Board
Best Air Program Award

CAROLINA, PUERTO RICO
Puerto Rico Aqueduct and Sewer Authority
Water Quality Award

CARTAGO, COSTA RICA
Chamber of Industries of Costa Rica and the Municipality of San Jose
Environmental Conservation Award

CARTAGO, COSTA RICA
Ministry of Natural Resources
Environmental Flag

CASTLEBAR/SWINFORD, IRELAND
Minister for Environmental Protection, Ireland
Good Environmental Management Award

CASTLEBAR/SWINFORD, IRELAND
European Better Environment Awards for Industry
Commendation Award in the Good Environmental
Management Category

ENVISION™ RECYCLING PROGRAM
I. V. Systems Division
Society of Plastics Engineers
First Annual Award for Recycling

HAYWARD, CALIFORNIA
City of Hayward
Earth Day Award

HAYWARD, CALIFORNIA
California Integrated Waste Management Board (CIWMB)
Waste Reduction Award

I. V. SYSTEMS
Illinois Environmental Protection Agency
Star Partner Award

IRVINE, CALIFORNIA (CVG)
City of Irvine
Environmental Excellence Award

IRVINE, CALIFORNIA (CVG)
California Integrated Waste Management Board (CIWMB)
Waste Reduction Award

IRVINE, CALIFORNIA (CVG)
National Awards Council for Environmental Sustainability
Environmental Achievement Certificate

MARICAO, PUERTO RICO
Puerto Rico Solid Waste Authority
Recycling Program Award

MONCTON, NEW BRUNSWICK, CANADA
Greater Moncton Chamber of Commerce
Park Beautification Award

MOUNTAIN HOME, ARKANSAS
Arkansas Environmental Federation
Hazardous Waste Minimization Award

MOUNTAIN HOME, ARKANSAS
Arkansas Recycling Coalition
Corporate Recycler of the Year

NORTH COVE, NORTH CAROLINA
North Carolina Recycling Association
Merit Award

OAKLAND, CALIFORNIA
CIWMB
Waste Reduction Award

POINTE CLAIRE, QUEBEC, CANADA
West Island Chamber of Commerce (Montreal)
Award for Environmental Achievement

RIVERSIDE, CALIFORNIA
Riverside County
Recycling Award

THETFORD, ENGLAND
Beazer Homes
Environment Award

WINNIPEG, MANITOBA, CANADA
Manitoba Round Table of Environment & Economy
Sustainable Development Award of Excellence

1995

ALLISTON, ONTARIO, CANADA
Baxter Canada
Regional Award for Pollution Prevention

AÑUSCO, PUERTO RICO
Puerto Rico Solid Waste Management Authority
Recycling

BAXTER
Lake County Youth Conservation Corps
Community, Careers, Conservation Award

BAXTER
Illinois Parks and Recreation Association and Illinois Association
of Park Districts
Community Service Award for the Advancement of Parks

CARTAGO, COSTA RICA
Ministry of Natural Resources
Ecological Flag

HAYWARD, CALIFORNIA
California Integrated Waste Management Board (CIWMB)
Waste Reduction Award

IRVINE, CALIFORNIA (CVG)
City of Irvine
Environmental Community Outreach Award

IRVINE, CALIFORNIA (CVG)
City of Irvine
Water Conservation Award

IRVINE, CALIFORNIA (CVG)
Mayor of Irvine
Waste Reduction Commendation

IRVINE, CALIFORNIA (CVG)
Mayor of Irvine
Arbor Day Tree Planting Commendation

IRVINE, CALIFORNIA (CVG)
Mayor of Irvine
Environmental Community Outreach Award

IRVINE, CALIFORNIA (CVG)
Harbor, Parks and Beaches Association
Park Cleanup Recognition

IRVINE, CALIFORNIA (CVG)
California State Legislature
Environmental Award

IRVINE, CALIFORNIA (CVG)
U.S. Congressmen
Environmental Community Outreach Recognition

IRVINE, CALIFORNIA (CVG)
Renew America
Environmental Sustainability Award

IRVINE, CALIFORNIA (CVG)
California Integrated Waste Management Board (CIWMB)
Waste Reduction Award

LOS ANGELES, CALIFORNIA
Catalina Harbor Clean-up
Golden Flipper Award

MCGAW PARK, ILLINOIS
Lake County Solid Waste Board
Award for Environmental Education

MISSISSAUGA, ONTARIO, CANADA
City of Mississauga
Best Environmental Program Award

MISSISSAUGA, ONTARIO, CANADA
Region of Peel
Outstanding Industrial and Commercial Environmental
Initiatives and Waste Reductions

MOUNT PEARL, NEWFOUNDLAND, CANADA
Government of Newfoundland
Environmental Award

MOUNTAIN HOME, ARKANSAS
State of Arkansas
Pollution Prevention Award

VERIE SANDBORG
National Association for Environmental Management
Certificate of Recognition

CURTIS STEPHAN
National Association for Environmental Management
Environmental Excellence Award

1996

AIBONITO, PUERTO RICO
Solid Waste Authority of Puerto Rico
Recycling Program Award

ALLISTON, ONTARIO, CANADA
Recycling Council of Ontario
Waste Minimization Award

ALLISTON, ONTARIO, CANADA
Arbor Committee of New Tecumseth
Corporate Green Challenge Winner

BAXTER
Lake County Youth Conservation Corps
Support Recognition

BAXTER
U.S. Environmental Protection Agency
Certificate of Appreciation for Voluntary Air Toxics Reduction

BAXTER
U.S. Environmental Protection Agency
WasteWi$e Sustained Leadership in Waste Prevention
Recognition

CARTAGO, COSTA RICA
Ministry of Natural Resources
Ecological Flag AA (2-Year Winner)

EDMONTON, ALBERTA, CANADA
Pollution Probe Foundation
Commuter Challenge Most Participation Award

ENVIRONMENTAL PERFORMANCE REPORT
Chicago Chapter of the International Association of Business
Communicators
Spectra Award of Excellence

ENVIRONMENTAL PERFORMANCE REPORT
District 4 of the International Association of Business
Communicators
Silver Quill Award of Excellence

ENVIRONMENTAL PERFORMANCE REPORT
Healthcare Marketing Report
Healthcare Advertising Merit Award

RAFAEL GUZMAN
Santiago de Cali University
Distinguished University Teacher (for teaching environmental
mangement at the university)

HARVARD, CALIFORNIA
California Integrated Waste Management Board (CIWMB)
Waste Reduction Award

ROUND LAKE, ILLINOIS
(I. V. SYSTEMS DRUG DELIVERY)
Governor of Illinois
Pollution Prevention Award

IRVINE, CALIFORNIA (CVG)
CIWMB
Waste Reduction Awards Program (4-Year Winner)

IRVINE, CALIFORNIA (CVG)
Renew America
Environmental Sustainability Award

LESSINES, BELGIUM
Belgium Federation of Chemical Industry
Responsible Care Special Recognition

MARSA, MALTA
US Environmental Protection Agency
Stratosphere Ozone Protection Award

MCGAW PARK, ILLINOIS
National Association for Environmental Management
Special Recognition Award for Environmental Excellence

MISSISSAUGA, ONTARIO, CANADA
Mississauga Board of Trade
Environmental User Award

THOUSAND OAKS, CALIFORNIA
City of Thousand Oaks
WasteWatch Gold Award for Outstanding Achievements in
Recycling

1997

BAXTER
Lake County Youth Conservation Corps
Recognition of Contributions

BAXTER
National Association of Physicians for the Environment
Inaugral Award for Business Activity to Protect Our
Environment and Health

CAROLINA, PUERTO RICO
Solid Waste Authority
Recycling Award

IRVINE, CALIFORNIA (CVG)
City of Irvine
Earth Day Environmental Outreach Award

IRVINE, CALIFORNIA (CVG)
RENEW America
Environmental Sustainability Award

ROUND LAKE, ILLINOIS (I. V. SYSTEMS)
Solid Waste Agency of Lake County
4R Award (for Waste Reduction)

Source: Baxter (1997b).

5 Beyond-compliance: findings and conclusions

This book examined environmental policymaking within two firms – Baxter International Inc. and Eli Lilly and Company – by exploring the internal processes that led them to selectively adopt Type 2 policies. Ten cases were examined: four common to these firms (underground tanks, 33/50, ISO 14000, and environmental audits), and one each specific to them (Responsible Care to Lilly and Green Products to Baxter).[1] All cases pertain to policymaking during 1975 to mid 1996. Cases were selected to ensure variation on independent variables (factors external and internal to firms). Further, as advised by King, Keohane, and Verba (1994), given the small sample size, I also ensured variation on the dependent variable (policy adoption or non-adoption). This discussion is summarized in table 5.1.

Firms' environmental policies were classified on two attributes: (1) whether they meet or exceed the requirements of laws and regulations, and (2) whether or not they meet or exceed the criteria specified in investment appraisal procedures. Based on this classification, four modal types of environmental policies were identified: Type 1 (those which go beyond-compliance and also meet or exceed the profit criteria), Type 2 (those which go beyond-compliance but cannot or do not meet the profit criteria), Type 3 (those which are required by law and also meet or exceed the profit criteria) and Type 4 (those which are required by law but cannot or do not meet the profit criteria).

Stringent monitoring and enforcement of environmental laws in industrialized countries, particularly the US, has ensured that managers have few incentives to systemically violate these laws. Consequently, the book focused only on beyond-compliance policies examining why firms selectively adopt Type 2 policies. The neoclassical economic theory suggests that firms' policies directly correspond to external stimuli, whether in the form of governmental regulations or market signals. It predicts that firms

---

[1] In this chapter, internal and external audits as well as Phase I and Phase II of Responsible Care are analyzed separately.

Table 5.1. *Findings*

	Baxter	Lilly
Underground tanks	Adopted: leadership	Adopted: leadership
EPA's 33/50	Adopted: leadership	Adopted: leadership
Responsible Care	n.a.	Initially not adopted; Finally adopted: leadership
ISO 14000	Not adopted	Not adopted
Internal audits	Adopted: power	Adopted: leadership
External audits	Adopted: power	Not adopted
Green products	Adopted: leadership	n.a.

would only adopt those policies that are either required by laws and regulations (Type 3 and Type 4) or that can be demonstrated *ex ante* as being profitable (Type 1).

The neoclassical theory is less useful in explaining why firms selectively adopt Type 2 policies. Sociological institutional theory and stakeholder theory are also inadequate to explain the research puzzle, although they contribute to understanding why firms adopt policies that may not deliver short-term or even long-term quantifiable profits. This is because focusing *solely* on how external factors shape firms' incentives (as the above theories do), *though necessary, is insufficient* to explain this puzzle. For example, only 13 percent of the eligible US firms adopted the Environmental Protection Agency's (EPA's) 33/50 program (chapter 4). Even among the top 600 companies, only 64 percent joined this program (Sarokin 1999). Any theory or perspective that focuses on the coercive character of the EPA (an external factor) as the sole explanatory variable is, therefore, under-specified to explain variations in firms' responses to the 33/50 program. Being under-specified means that as a part of its causal explanation, the neoclassical theory leaves out one or more of the important independent variables; in the context of beyond-compliance policies, internal processes of firms constitute the omitted variables.

Given our inability to unbundle firms by employing the neoclassical *theory*, the institutional *theory*, and the stakeholder *theory*, I turned to the new-institutionalist *framework*.[2] This was also an ontological departure since managers within firms, and not firms themselves, became my unit of analysis. Both new-institutionalists and neoclassicalists assume methodological individualism. However, by treating firms as unitary actors, neoclassicalists make no distinction between firms and individual actors who

---

[2] The distinction between a theory and a framework is discussed subsequently.

work within these firms. Since many scholars of organizational theory (including those subscribing to the behavioral theory of the firm) view firms as composite actors, a strategy of unpacking firms is not unique to new institutionalists.

Next, I explained the adoption or non-adoption of Type 2 policies based on preferences (policy supporters versus policy skeptics) and endowments of actors (specifically, their position in a firm's hierarchy) as articulated within: (1) internal institutions and structures of their firm, and (2) the external environment in which their firm functioned. My strategy therefore had three steps. First, I opened up the "black-box" called the firm to identify the main actors in the firm's environmental policymaking, their preferences for Type 2 policies, their position in the hierarchy, and the firm's internal procedures and structures. Second, I linked actors, procedures, and structures to processes of environmental policymaking. Processes imply two things: (a) the criteria employed by managers for assessing the profitability of projects, including environmental projects and (b) the way in which managers interpret these criteria to support their desired projects, and, as a result, how projects may not always be subjected to such formal appraisal procedures. Third, since firms constantly negotiate with the external world, I examined how policy supporters, successfully or unsuccessfully, invoked such external factors to influence internal processes.

New institutionalists treat policies of firms as institutions, and firms themselves as organizations. Hence the book examined the adoption or non-adoption of a specific category of institutions: Type 2 policies. New institutionalists focus on two sets of questions: first, how do institutions evolve in response to individual preferences, strategies, and endowments, and, second, how do they affect policy outcomes? This study focused on the first question by examining the adoption and non-adoption of a firm-level institution: Type 2 policies.

New institutionalists assume methodological individualism, bounded rationality, and pursuit of self-interest. Given their bounded rationality, managers employ established procedures such as capital budgeting to assess *ex ante* the profitability of projects, thereby attempting to maximize shareholders' wealth. In capital budgeting, managers project future cash flows emanating from a project and discount them by the firm's cost-of-capital. Cash flow projections are the best managerial estimates of an uncertain future. To convert issues of uncertainty into issues of risk, managers weight the cash flows by their subjective probabilities. They periodically update probabilities in light of market or regulatory changes. Needless to say, managers' assessments of the future vary substantially. To map out variations in projected profit and cash flows, managers subject these estimates to sensitivity analysis. Thus, managers adopt standard operating procedures to cope with uncertainty and to establish transparency and impartiality in the project appraisal process. A project is *ex ante* profitable if it meets the criteria of such established procedures. Importantly, *ex ante* profitable projects may result in *ex post* losses if market or regulatory conditions change in unanticipated ways.

There are other methods of investment analysis as well, such as full-cost accounting, and life-cycle analysis that firms could potentially employ. These methods force firms to take into account long-term social (as well as private) costs, thereby internalizing environmental externalities. I did not examine these alternative investment appraisal techniques for several reasons. First, Lilly and Baxter do not routinely employ any of these methods for appraising projects (while they do employ capital budgeting). Further, even as a non-routine measure, they did not employ them in making decisions on Type 2 policies examined in chapter 4. Second, since the methodologies and operationalization of these techniques is not standardized, they lack legitimacy, especially with the finance and accounting managers. Finally, these techniques may not be compatible with the dominant paradigm that the main objective of firms is to maximize shareholders' wealth, an objective that capital budgeting focuses on.

I argued that certain managers may think that a given Type 2 policy is beneficial for their firm in the long-run even though its impact on profits cannot be quantified. Clearly, such policies cannot satisfy the requirements of established procedures of investment analysis. These managers then face a difficult task of convincing skeptics that this policy should not be subjected to such procedures. In these situations the internal policy-making dynamics become important. Policy supporters *do not always* succeed (as in ISO 14000 and external audits in Lilly). If they do, it reflects power-based or leadership-based processes.

Power-based theories view firms as representing the domination of one set of actors over another. They, therefore, predict that Type 2 policies will be adopted only if they have powerful sponsors such as hierarchically superior managers.[3] Some of Baxter's managers opposed inviting Arthur D. Little (ADL) to help them define state-of-the-art environmental standards and to audit Baxter's environmental programs against these standards (chapter 4). Interventions by very senior managers including

---

[3] Theoretically, there is another policy category: Type 1 policies that were not adopted due to the opposition from powerful actors. This theoretical category has little practical relevance since senior managers, whose compensation is often linked to quantifiable profits and the firm's stock price, will have few incentives to oppose such policies. I could not find instances of such opposition in Baxter or Lilly.

Baxter's Chief Executive Officer, ensured that such objections were over-ruled. Consequently, this policy was adopted by imposition.

Leadership-based theories suggest that power-based (and transaction cost) explanations cannot fully explain the nature of firms. They argue that firms emerge only through the intervention of leaders who can convince other managers to reassess their assumptions and preferences regarding the costs and benefits of collective action. Policies carry leaders' imprints in that firms would not have adopted them in the absence of leaders' interventions. Type 2 policies on underground tanks, the 33/50 program, and Responsible Care were initially opposed by some managers since they do not meet the formal profit criteria. Over time, however, policy supporters succeed in convincing policy skeptics that these policies indeed serve the long-term interests of their firms, although their profit contributions cannot be quantified. Such policies are eventually adopted by inducing consensus.

Ex ante, profits are mere projections. Of course, by considering alternative scenarios (through sensitivity analysis), managers endeavor to make more rigorous projections, a task that has been greatly facilitated by computer packages. In some instances, sole reliance on formal assessment procedures is not useful since it is not possible to quantify the profit impact. This creates a political space for "discursive struggles" within firms and opportunities for managers who claim to have a "vision" to push through their pet projects. These policies are implemented if either such managers are able to convince others that such projects help the firm in the long-run (leadership-based processes) or if they have the authority to impose their "vision" on others (power-based processes). Although the policy supporters often claim increases in long-term profits, they provide no estimates. This suggests that in some instances profit no longer remains an "objective" concept whose assessment is invariant across actors. I am not arguing that established methods of project appraisal are irrelevant. They are indeed relevant and that is why it is difficult for the supporters of Type 2 policies to justify why their pet policy should not be subjected to established project appraisal procedures. Such exceptions do occur, especially in the realm of social policies such as environmental policies. This book attempted to understand the processes that lead to such exceptions and the conditions under which they occur.

To summarize, neoclassical economics predict that managers employ some "objective" criteria to assess the profitability of policies. Profitability would be assessed *ex ante*, and only policies meeting or exceeding this criteria would be adopted. In the context of Type 2 policies, managers cannot or do not employ such well-defined criteria. For example, it is difficult to assess the profitability of environmental audits that do not require up-front capital expenditure and that do not generate revenues (ISO 14000). Even in some circumstances (such as underground storage tanks and the 33/50 program) that involved significant capital expenditures, policy supporters succeeded in not employing established appraisal procedures.

This study has important implications for environmental policy debate. Many argue that firms are often oblivious of the low-hanging fruit – "win–win" projects that generate profits and deliver superior environmental performance (Porter 1991; Porter and van der Linde 1995). Hence, stringent environmental laws may force firms to harvest this low-hanging fruit, thereby benefitting themselves and also the natural environment. Win–win strategies assume that firms can earn quantifiable profits by adopting environmentally progressive policies. Some projects (Type 1 policies) certainly fall in this category, especially the ones that lead to pollution prevention or reduction. Type 2 policies such as adopting new management systems can seldom be demonstrated as being *ex ante* profitable. A focus on win–win projects as a pillar of environmental policy is misplaced. Many "desirable" environmental policies could conceivably benefit firms in the long run only and it is often difficult to justify on traditional economic grounds. The win–win rhetoric creates false expectations that lead to backlash when not met. The emergence of the so-called "green-wall" in the mid 1990s (chapter 4) was primarily due to the managerial skepticism about the economic soundness of environmental policies. Even if low-hanging fruits exist, many of them have been already harvested by firms in response to public pressure, new managerial consciousness, and stringent regulations. Now, firms are struggling to harvest the high-hanging fruits or fruits from some other tree, and these may not constitute a win–win scenario. This book, therefore, argues that the new genre of environmental policies and regulations need to be justified on non-economic grounds as well because their economic rationale may be difficult to demonstrate.

Theoretical implications

Research objectives often fall in the realm of either theory testing or theory building. Theory testing is useful when the existing theoretical tools are sufficient to examine a given puzzle. Of course, in the process of testing existing theories, researchers may suggest modifying them. Theory building is useful when existing theoretical tools are insufficient in explaining a given research question. For example, I classify Coase's (1937) landmark article, "The Nature of the Firm," as being in the realm of theory building since the neoclassical economic theory did not sufficiently explain the major theoretical question: why do firms arise in the first place? This book could also be viewed in the realm of theory building. Existing theories – neoclassical economics, sociological institutional theory, and stakeholder theory – inadequately explain why firms selectively adopt Type 2 policies. I have, therefore, employed two theories that together explain this research question. However, this is only an initial step in developing better theory to understand this question. I have attempted to understand the notion of efficiency by differentiating between substantive efficiency and procedural efficiency, and argued that to examine intra-firm dynamics, procedural efficiency is more appropriate. As a second-round effect, substantive efficiency is predicted to influence procedural efficiency. However, this impact is mediated by a variety of institutions such as competition in the market as well as managerial abilities to learn from such market signals.

This book has three theoretical implications. First, ontologically, it makes an argument in favor of methodological individualism: the policies of composite actors should eventually be traced to preferences, endowments, and strategies of individual actors, as articulated within their internal institutions. Second, many social phenomena such as corporate responses to beyond-compliance policies cannot be adequately understood by employing one theory; rather this requires employing multiple theories and a common framework to link them. Third, it makes an argument for "bringing back leadership" to political economy, since leaders often play crucial roles in institutional evolution and change. I elaborate on these implications below.

Composite actors

The neoclassical theory (as well as sociological institutional and stakeholder theories) treat firms as undifferentiated unitary actors responding only to external factors. This spartan view of the firm serves well to predict firms' responses to Type 3 and Type 4 policies (both are required by law) and Type 1 policies (those that meet or exceed the requirements of capital budgeting) but not Type 2 policies.

How can the selective adoption of Type 2 policies be explained? There are two strategies, both sacrificing parsimony to increase the explanatory power of the theory. The first strategy is to view firms as differentiated unitary actors and not as undifferentiated unitary actors. With this additional assumption, one could argue that differentiation across firms makes them respond differently to the same external stimulus. The second strategy is to treat firms as differentiated composite actors and focus on their internal processes.[4] One could then argue that the differences in internal processes explain variations in response to Type 2 policies. I elaborate on these two strategies below.

In employing the first strategy, firms could be differentiated on the basis of a variety of indicators such as assets and sales (large firms versus small firms) or industry type (engineering firms, chemical firms, electronic firms, etc.). As discussed in chapter 1, UNCTAD's benchmark survey on MNEs' environmental practices suggests that MNEs' environmental programs are a function of three factors: size of the firm, nationality of the parent firm, and industry type (UNCTAD 1993). It reports that larger MNEs (with sales above $4.9 billion) have stronger environmental programs than smaller MNEs.

A strategy of treating firms as differentiated unitary actors is considerably superior to one that treats them as undifferentiated unitary actors. Such differentiation based on their external attributes enhances the ability of theory in explaining why firms with given attributes tend to adopt specific kinds of environmental programs. In effect, by diluting parsimony, our theoretical tools have gained superior explanatory and predictive power.

We still cannot explain why, in response to the same external stimulus, firms sharing similar attributes adopt different policies. For example, while Baxter and Lilly share characteristics identified in the UNCTAD survey, they have responded differently to the demands of environmental groups for transparency in their environmental programs. Baxter invites external auditors to evaluate its environmental programs whereas Lilly does not. Further, we are unable to explain why a given firm responds differently to similar external stimuli. For example, Lilly has responded differently to the adoption of industrial codes of practices; it has adopted Responsible Care but not ISO 14000.[5]

What is the next step? One strategy could be to make the classification of firms' attributes more complex. However, there is no theoretical reason why this additional complexity will improve the theory's explanatory power. I am of the view that any classification must have a theoretical basis: we must have a logic to classify in the manner that we do. Though one could observe patterns and infer theoretical reasons why

---

[4] There is a third strategy as well: firms as undifferentiated composite actors consisting of employees with identical endowments and preferences and functioning in identical sets of internal institutions. Clearly, this category has little practical relevance. It is also not a theoretically useful category since the objective of unpacking firms is to identify differences in their internal structures and processes, and link these differences to variations in policy outcomes.

[5] For an elaboration of this argument, see Prakash (1999a).

such patterns arise, such explanations are under-specified in that there could be multiple reasons for the same pattern. This book, therefore, does not adopt this approach.

Perhaps, a paradigmatic shift is required to change the focus of inquiry from firms' external factors to their internal ones. This requires employing the second strategy of unbundling firms; treating them as differentiated composite actors, and not as differentiated unitary actors. This calls for revising the ontological assumptions; instead of treating firms as units of analysis, the focus shifts to individual managers within firms as units of analysis.

An examination of inter-manager interactions, particularly between policy supporters and policy skeptics, should explain the following: (1) in response to the same external stimulus (such as stakeholder scrutiny), why some firms adopt certain Type 2 policies, while others do not (only Baxter invites external auditors); (2) in response to similar external stimuli (for example, industrial codes of practice), why the same firm adopts different policies (Lilly adopts Responsible Care but not ISO 14000). Such inter-manager interactions could be dissimilar within and across firms due to factors such as: (1) variations in firms' internal decision-making institutions; (2) levels of opposition of policy skeptics (specifically, the "losers" from changes in collective-choice-level institutions); (3) the hierarchical power of policy supporters in relation to policy skeptics; (4) the credibility and persuasive abilities of policy supporters in rallying policy skeptics; and (5) the managerial perceptions of the coercive power of external institutions that are encouraging or discouraging adoption of a given Type 2 policy.

Multiple theories and common framework

The second theoretical implication of this book is that explaining social phenomena (Type 2 policies in this book) often requires multiple theories linked by a common framework. The distinction between frameworks and theories is emphasized in the literature on the Institutional Analysis and Development framework developed at the Workshop in Political Theory and Policy Analysis (Ostrom 1990; Ostrom, Gardner, and Walker 1994). According to Elinor Ostrom:

A framework helps to identify the elements and relationships among these elements that one needs to consider for institutional analysis . . . they organize diagnostic and prescriptive enquiry . . . provide a metatheoretical language to compare theories . . . attempt to identify universal elements that any theory relevant to the same kind of phenomena would need to include. (1996: 4)

In contrast, theories help analysts to specify elements of a framework that are relevant to a given question. They make specific assumptions to diagnose this question, explain its sources, and formulate some solutions. Several theories are usually compatible with any framework (Ostrom 1996). A given theory may be compatible with other frameworks as well. For example, power theories are compatible with the Marxian framework. This book employs two sets of theories linked together within a new-institutionalist framework.[6] Both these theories view institutions as potential artifacts. However, they differ in their explanations of how and why institutions emerge. No theory alone can explain all aspects of my research question.

Neoclassical economics predict that Type 2 policies would not be adopted. Power-based theories predict that Type 2 policies would be adopted if policy supporters have sufficient coercive power to over-rule policy skeptics. Thus, such policies may be adopted without changes in dissenters' preferences. I disagree with scholars who suggest that power-based explanations are preeminent, and leadership is only a subtle exercise of power. Later in the chapter, along with this issue, I discuss my strategy to infer whether or not policy skeptics have changed their preferences.

In contrast to power-based processes, leadership-based policies involve changes in preferences. This requires an assessment of preference intensity. My strategy has been to examine the extent of organizational change, thereby assessing the perceived "loss" of "losers" from such changes. If the loss is significant, losers can be predicted to exhibit strong preferences for not adopting Type 2 policies, and would therefore have few incentives to revise their preferences. As a result, if such policies are implemented at all, it would be through power-based processes.

Leadership-based policies bear the imprints of particular leaders who employ their persuasive skills to convince policy skeptics of the long-term payoffs of adopting Type 2 policies. Over time, in some instances, policy skeptics change their preferences, and such policies are then adopted consensually.

Do these theories constitute alternative explanations; that is, do they suggest different logics for understanding a given phenomenon? If so, do we need to prioritize these theories? Prioritizing can have two meanings.

---

[6] In the International Business literature, Dunning (1993) employs the eclectic paradigm – the OLI framework – to link three sets of theories that together explain why firms invest abroad. These theories focus on three sets of variables that pertain to the competencies of *organizations*, the advantages for firms to *locate* abroad, and the need to *internalize* transactions within the firm instead of using market processes.

First, one could rank these theories on their explanatory and predictive powers. This is not done since I see them as working *together* to explain a broader puzzle. Second, one could prioritize these theories in terms of the research chronology. This study followed a particular format in my analysis of the ten cases. First, it employed neoclassical economics for understanding the adoption or non-adoption of a particular policy. This is because it is generally assumed that the preeminent objective of firms is pursuit of profits. Since neoclassical economics could not explain the selective adoption of Type 2 policies, the study employed leadership-based and power-based explanations.

Since both power-based and leadership-based perspectives add to our understanding of social phenomena, why should there be an insistence that one perspective must always prevail?[7] Why not view these perspectives as complementary that together help us to understand complex social reality? In employing three models (rational actor, organizational process, and governmental politics) to examine the central puzzles of the Cuban missile crisis, Allison notes:

> The three models are obviously not exclusive alternatives. Indeed, the paradigms highlight the partial emphasis of each framework – what each magnifies and what it leaves out. Each concentrates on one class of variables, in effect, relegating other important factors to a *ceteris paribus* clause. The models can therefore be understood as building blocks in a larger model of the determinants of outcomes. (1971: 275)

The distinction between a framework and a theory provides a persuasive rationale for employing multiple theories. Theories and frameworks constitute two different levels of theoretical inquiry. Often by focusing on the incorrect level of analysis (the level of theory) we find ourselves in midst of unnecessary contestations on the superiority of one theory over another. An understanding of why firms adopt or do not adopt beyond-compliance policies would have been incomplete had I employed only leadership-based or power-based theories. On this count, new institutionalism is particularly attractive in that it allows multiple perspectives to flourish, thereby encouraging scholars in the increasingly divided discipline of political economy to find commonalities. Of course, many scholars disagree with the basic assumptions of new institutionalism. Some consider it equivalent to rational-choice institutionalism. Rationality implies that human behavior is consequence governed, and that actors estimate their costs-benefits prior to action. In contrast, many scholars

---

[7] In this context see Weimer (1997) where multiple theories have been employed (economics, public choice, and distributional) to have better specified understanding of the privatization experiences in Eastern Europe and the former Soviet Union.

view human behavior to be rule governed (March and Olsen 1989; Sandholtz 1999). Still, others give primacy to ideas and discourse in shaping human behavior (Hall 1986; Wendt 1992). For them, ideas are not mere intervening variables since preferences are socially constructed.

Leaders matter

New institutionalists assume methodological individualism implying that all actors are ontologically equal; it does not imply that all individuals are alike. Individuals differ in their preferences, their access to resources, their abilities to employ such resources, and their capacities to influence outcomes. Some individuals have greater capacities and willingness to influence institutional evolution in favor of Type 2 or other policies. The discussions on underground storage tanks, the 33/50 program, and Responsible Care suggested that such leaders demonstrate capacities to convince policy skeptics to revise their assessments of the long-term benefits of Type 2 policies.

This study does not view power-based processes reflective of leadership. Structurally advantaged individuals such as hierarchical superiors may force the adoption of Type 2 policies but such interventions do not lead policy skeptics to revise their assessments of Type 2 policies. Even recently I observed continued opposition in Baxter to external audits by Arthur D. Little (ADL). The ADL consultant was more anxious about his presentation to the Divisional Environment Managers (DEMs; policy skeptics) than to the Environmental Review Board which is the highest decision-making body on environmental issues. After attending his presentation to the DEMs, I better appreciated his reasons for being anxious; some DEMs continue to oppose external audits by ADL. Hence, I infer that although external audits have been mandated by senior management, many DEMs have not revised their preferences against them.

Similarly, though a senior manager held up the implementation of the Community Outreach Program under the aegis of Responsible Care in Lilly, policy supporters did not give up their efforts. After about eighteen months, once the policy skeptic retired, policy supporters persuaded Lilly to implement community outreach programs.

A focus on leadership is also helpful in strengthening the dialogue between political economy and organizational theory. Though both political economists and organizational theorists study collective action, their intellectual agendas appear to have drifted apart. Political economists tend to study the dynamics between politics and economics. A study of politics is often (incorrectly) equated to the study of governments, and a study of economics to the study of markets. Consequently, the intellectual arena for many political economists is limited to studying two institution types: markets and governments. Some political economists, especially public choice scholars, tend to employ the tools of economics to study politics (Mueller 1989). On the other hand, organizational theorists tend to employ the tools of sociology. Sociologists also tend to focus less on markets. Due to differences in methodological tools and research focus, the level and quality of dialogue between economists and sociologists leaves much to be desired. This drift is unfortunate since both political economists and organizational theorists study collective action; they have much to contribute to each other. New institutionalism, especially the Ostrom version, combines insights (at least) from political science, economics, and sociology. Therefore, it has the potential to initiate a dialogue between organizational theory and political economy. Since leadership is a highly researched subject in organizational theory, this is one area where such cross pollination can begin.

Does studying leadership imply that our research will become focused on examining preferences and strategies of leaders, and not of "ordinary folks"? If so, will it distort our understanding of social processes? Such debates have reverberated in other contexts as well. For example, historians such as Ambrose (1967) emphasize the role of leaders, while others such as Zinn (1995) and Appleby, Hunt, and Jacob (1994) emphasize the contributions of popular movements.[8] In international relations, realists such as Morgenthau (1978) and Kissinger (1964) emphasize the role of leaders in shaping foreign policy. In their study of the Indian Freedom movement, historians such as Majumdar, Raychaudhuri, and Datta (1958) focus on the contributions of Gandhi, Nehru, Bose, and other stalwarts, while "subaltern historians" emphasize the role of popular movements (Sarkar 1989).

Clearly, for studying social processes at the firm level or country level, one needs to examine the contributions of both leaders and "ordinary folks," as well as the structures within which they function. Political economists focus on structures (with their emphasis on institutions) as well as "ordinary folks" (an implication of methodological individualism). Some scholars also examine the roles of structurally advantaged actors in shaping social processes and I have termed such explanations power-based theories. However, leadership seems to be a relatively neglected area in the study of political economy.

Would introducing leadership as a key variable in political economy lead to elitist explanations of social processes? Such concerns are unwarranted as we are interested in understanding the phenomenon of leader-

---

[8] I thank Cynthia Yaudes for this point.

ship, how it affects collective action, and the evolution of institutions. There is no assertion that only certain types of people can display leadership. What is important is that leadership may influence the organization and form of collective action, whether in small self-governing communities or in global corporations.

Policy implications

This book examined an important policy issue concerning firms' responses to environmental issues. It is often believed that firms are the main agents of environmental degradation: firms cut corners to save on environmental costs, and, as a result, there is an inherent conflict between environmental sustainability and firms' objectives. This study challenges the simplistic notions that suggest that firms will invariably adopt policies that minimize the costs borne by them.

Two policy implications flow from this book. First, factors external to firms (and these could be influenced by policymakers) may have a critical, though *not a deterministic*, influence on environmental policy processes within firms. Second, firms often adopt those Type 2 policies that require fewer changes in their internal organizations and institutions, especially at the collective-choice level. This has important implications for policy design, again a variable that could be manipulated by policymakers.

The role of external factors

In contrast to the neoclassical economic theory (as well as sociological institutional and stakeholder theories) that places *exclusive* emphasis on external factors, this study emphasizes the importance of *both* internal processes and external factors in influencing whether or not firms adopt Type 2 policies. Clearly, external factors are important since they influence the "rules" and "norms" (Crawford and Ostrom 1995) within which firms conduct their business. They may also directly influence managerial behavior. For example, individual managers in US firms can face civil and criminal prosecution for environmental violations by their firms. However, managers may not always take the external institutions as given. As the literature on regulation points out, sometimes managers may even shape them. Stigler (1971) argues that firms may lobby for regulations in certain circumstances. Bernstein (1955) suggests that business firms may "capture" governmental agencies that are supposed to regulate them and Kolko (1963) provides an interesting account of how the railroad industry "captured" the regulatory agencies.

Previously, environmental policies were classified as compliance driven (Type 3 and Type 4) or as beyond-compliance (Type 1 and Type 2). External institutions (laws and regulations in our case) alone are sufficient to predict firms' responses to compliance driven policies. However, external factors may also influence adoption or non-adoption of Type 2 policies by impacting firms' internal policy processes. In particular, three attributes of the external environment are important: "power," "legitimacy," and "urgency" (Mitchell, Agle, and Wood 1997) in the context of a given policy. External stakeholders are powerful if they can compel managers to do something that the managers would not have done otherwise. Legitimacy implies that managers view the actions or demands of the external stakeholder as appropriate within a given set of norms and beliefs. The urgency dimension suggests that managers believe that there is a need for immediate action in response to the demands or requests of the external stakeholders. Managerial perceptions of the three attributes can be expected to vary within and across policies. With Type 2 policies creating a political space for "discursive struggles," policy supporters and policy skeptics compete to sway managerial perceptions one way or other.

As suggested in table 5.2, if managers perceive external factors which promote a given Type 2 policy as legitimate, urgent, and significant in shaping their firm's market and non-market environment, policy supporters have greater *incentives* to mobilize internal coalitions. Since efforts of policy supporters have greater *credibility* among policy skeptics, there is a greater likelihood that internal coalitions in support of a given Type 2 policy will emerge.

Of the six Baxter's policies the book examined, five were adopted (underground tanks, green products, 33/50, internal, and external audits). In three of the five adopted policies, Baxter's managers perceived external actors encouraging policy adoption as important (underground tanks, green products, and 33/50). For internal and external audits, however, they perceived external factors as important but discouraging policy adoption. Baxter did not adopt ISO 14000, primarily because managers considered external factors encouraging this policy to be relatively unimportant. Also, policy supporters could not capture the top management and have it mandate policy adoption.

Similarly, of the six Lilly's policies the book examined, four were adopted (underground tanks, 33/50, phase two of Responsible Care, and internal audits). In three of these four policies, managers perceived external factors encouraging policy adoption as significant (underground tanks, 33/50 and phase two of Responsible Care). For internal audits, however, managers viewed external factors as important but discouraging

Table 5.2. *External factors*

Policy	Outcome/Process	Importance of external factors
Baxter		
Underground tanks	Adopted: leadership	High; positive
EPA's 33/50	Adopted: leadership	High; positive
Responsible Care	n.a.	n.a.
ISO 14000	Not adopted	Low; positive
Internal audits	Adopted: Power	High; negative
External audits	Adopted: Power	High; negative
Green products	Adopted: leadership	High; positive
Lilly		
Underground Tanks	Adopted: leadership	High; positive
EPA's 33/50	Adopted: leadership	High; positive
Responsible Care	Initially not adopted; Finally adopted: leadership	High; positive
ISO 14000	Not adopted	Low; positive
Internal audits	Adopted: leadership	High; negative
External audits	Not adopted	High; negative
Green products	n.a.	

Summary

Importance of external factors	*Adopted*	*Not adopted*
High and positive	Underground tanks (Baxter, Lilly) 33/50 (Baxter, Lilly) Responsible Care (Lilly) Green products (Baxter)	Responsible Care (Lilly)
High and negative	Internal audits (Baxter, Lilly) External audits (Lilly)	External Audits (Baxter)
Low and positive		ISO 14000 (Baxter, Lilly)

policy adoption. Similar to Baxter, Lilly did not adopt ISO 14000 since managers considered external factors as relatively unimportant. Further, policy supporters could not influence the top management and have it mandate policy adoption.

Both Lilly and Baxter adopted the 33/50 program primarily because its sponsor, the EPA, had the capacity to significantly influence these firms' business environments. The EPA was also perceived as a legitimate actor and the need to subscribe to 33/50 program was crucial to retain public confidence. To recap, 33/50 aimed to reduce emissions of TRI chemicals. The TRI database was accessible to the public, and environmental groups used it to identify key polluters in a given county and state. The EPA's sponsorship of this program facilitated the emergence of internal coalitions (leadership-based policy). Even though 33/50 was a voluntary program, its supporters within Baxter and Lilly argued that its adoption was necessary for their firms to remain in the good books of the EPA. The perception among managers of the coercive character of the EPA created incentives for and imparted credibility to the actions of policy supporters to rally policy skeptics. Policy supporters invoked their firms' experiences with the EPA in the late 1980s and succeeded in convincing the policy skeptics that adopting 33/50 program would create substantial goodwill for their firm within the EPA as well as with other external stakeholders. However, not every US firm perceived the EPA as being sufficiently coercive and/or legitimate, and few believed that adopting 33/50 would keep their firms in the good books of the EPA. As a result, only 13 percent of firms releasing TRI chemicals adopted the 33/50 program.

Leadership-based processes also explain Lilly's and Baxter's adoption of Type 2 policies on replacing underground tanks. Policy supporters within these firms stressed that even a single leak from an underground tank could seriously harm their firm's reputation. Since both Lilly and Baxter are in the health-care business where a firm's reputation is very important, this could inflict substantial long-term (non-quantifiable) financial damage on them. In Lilly, the coercive character of the US legal system, particularly the award of punitive damages, was important in rallying support on this policy. Policy supporters also stressed that Lilly must ensure that communities in its facilities' vicinity view Lilly as a responsible and trustworthy corporate citizen. Nearly 90 percent of Lilly's tanks are located in its Tippecanoe and Clinton facilities in Indiana. Previously, there had been unsubstantiated complaints that leakages from underground tanks contaminated the region's aquifers. Hence, the local communities (external stakeholder) needed reassurance that Lilly's storage tanks did not leak; installing expensive above-the-ground tanks (Type 2 policy) enabled visitors to Lilly's facilities to visually inspect the storage tanks.

In part, Lilly's enthusiastic support for Responsible Care can be attributed to its desire to be a major player in the Chemical Manufacturers Association (CMA). Lilly is a show-case example of the successful implementation of Responsible Care. Since the CMA has an important voice on policies impacting the chemical industry, the major players within the CMA have opportunities to pursue their firms' agenda under its aegis.

Policy supporters highlighted the significant payoffs of adopting Responsible Care in terms of increased credibility with CMA member firms, thereby acquiring greater ability to influence CMA's agenda.

In the "green products" case (chapter 4), the book argued that the nature of the business portfolio (external factor) critically influences policies on the manufacturing and marketing of green products. Consequent to the EPA's proposal on air-quality standards and the rising costs of disposing wastes, Baxter's customers began talking about packaging reductions and reducing PVCs. This created an opportunity for supporters of green products to argue that Baxter has important business reasons to be proactive and to consider marketing products that offer environmental benefits as a distinctive product offering. In contrast, since Lilly manufactures and markets ethical drugs, its customers (doctors) potentially have little interest in the greenness of drugs; they look for attributes such as efficacy, cost of treatment, and availability. Hence, for doctors, the greenness of Lilly's products and corporate policies are "hygiene" factors (their absence may dissuade them from prescribing Lilly's products), but not "motivators" (their presence may not persuade them to prescribe Lilly's products).

This discussion has important implications for the environmental policy discourse since "greening" can occur at the level of the firm as well as the level of the product; and one may not always lead to the other. An important question therefore is: what kind of greening should be encouraged? For example, the General Agreement on Tariffs and Trade (GATT), now the World Trade Organization (WTO), allows countries to subject its imports to product standards but not process standards (Esty 1994). Thus, GATT creates incentives for firms to "green" their products but not their management systems. On the other hand, if ISO 14000 certification becomes a *de facto* requirement for exporting to the European Union, firms will have greater incentives to green their management systems.

Policy supporters face a difficult task if policy skeptics and the top management do not consider external factors to be important, legitimate, and urgent; particularly in terms of the external stakeholders' abilities to impose significant excludable costs or to offer significant excludable benefits. It impedes their efforts to mobilize coalitions or to "capture" the top management. Lilly's and Baxter's wait-and-see policies on ISO 14000 (at least, until mid 1996) were partially due to the non-coercive character of the International Organization for Standardization, the sponsor of ISO 14000 (the internal factor impeding the adoption of ISO 14000 being the emergence of a "green-wall"). The policy implication is that ISO 14000 must provide excludable benefits or impose excludable costs on firms and their managers. Only then can leadership-based or power-based processes be expected to empower policy supporters. For example, if the EPA lends visible support to ISO 14000 standards, perhaps US firms may consider ISO 14000 more seriously. The EPA's stand against granting attorney–client privilege to environmental audits (and ISO 14000 involves such audits), makes ISO 14000 even less attractive to US firms.

The discussion on environmental audits offered yet another perspective on the influence of external factors on intra-firm dynamics. As discussed in the case on environmental audits (chapter 4), many states have passed laws that encourage firms to undertake self audits. The EPA ostensibly supports self audits. However, it vehemently opposes granting attorney–client privilege to information gathered during such audits, thereby discouraging self audits. Given the coercive powers of the EPA, external factors appear to discourage firms from undertaking audits.

Both Lilly and Baxter have established internal audit programs. How does one explain this anomaly? Policy supporters successfully portrayed audits as tools to facilitate compliance and to improve environmental management systems. They contended that since their firms were committed to comply with or exceed the requirements of all applicable laws and regulations, the firm had nothing to fear from audits. And if there were instances of unintentional violations (the assumption being that there are not intentional systemic violations), firms would rather report them themselves, and correct them as soon as possible. Policy supporters perceived that, in the long-term, the EPA and other external actors that opposed such audits would realize that firms are really not misusing them, and will therefore begin to view audits more favorably. As a result, proactively establishing internal audit systems will have long-term payoffs.

Demythologizing beyond-compliance

An important finding is that if policy skeptics and top management perceive that external organizations encouraging a Type 2 policy can impose excludable costs or provide excludable benefits, they are more amenable to listening to policy supporters. The case studies described how the managerial perceptions of the EPA's coercive character enabled policy supporters to successfully argue for Type 2 policies. Adopting Type 2 policies appears to have been encouraged by existing regulatory regimes. Existing compliance standards set the base level upon which beyond-compliance initiatives take place.

Arguably, firms adopt beyond-compliance initiatives primarily to preempt even more stringent regulations or to shape future regulations.

To some extent this is true since many policy supporters argued that firms need to adopt Type 2 policies if they wish remain a credible player in the environmental policy discourse. However, one needs to be careful in attributing adoption of *all* Type 2 policies to such motivations. Firms face collective action dilemmas in adopting Type 2 policies; only concerted action by firms can potentially ward off tougher regulations or shape future regulations. The preemption argument may hold for industry-level initiatives such as Responsible Care and ISO 14000 that explicitly seek to preempt stringent regulations. By converting goodwill for firms among regulators and other external actors from a public good to a club good, industry-level initiatives mitigate collective action dilemmas (chapter 2). Consequently, firms have greater incentives to adopt such Type 2 poli-cies. It is, however, difficult to explain how Lilly's adoption of the most expensive route to replace underground tanks could ward off stringent regulations or how Baxter's use of external auditors could convince the EPA to adopt a more industry-friendly approach by granting attor-ney–client privileges on environmental audits.

Organizational change

Type 2 policies requiring significant organizational change are less likely to be adopted by leadership-based processes. By creating new institutions and structures, or by reallocating responsibilities within existing struc-tures, organizational changes upset the status-quo. Actors "losing" from such changes have incentives to oppose them. If the organizational changes to implement such a policy are "significant" (see below), it becomes difficult to pacify the losers without diluting the integrity of the proposed policy. As summarized in table 5.3 below, policies requiring significant changes may not be adopted by leadership-based processes since it is difficult to induce losers to cooperate. If they are adopted at all, they reflect power-based processes whereby structurally powerful policy supporters were able to impose their will on policy skeptics. Though policy skeptics remain dissatisfied, they are forced to accept the new policy.

Of the six Baxter's policies the book examined, organizational changes were significant only for environmental audits – both internal and exter-nal. Not surprisingly, Type 2 policies on internal and external audits were adopted by power-based processes. Organizational changes were less significant for the other four, three of which were adopted by leadership-based processes (underground tanks, 33/50, and green products), and one which was not adopted (ISO 14000).

Of the six Lilly's policies examined, organizational changes were

Table 5.3. *Organizational change*

	Outcome/Process	Organizational change
Baxter		
Underground tanks	Adopted: leadership	Low
EPA's 33/50	Adopted: leadership	Low
Responsible Care	n.a.	n.a.
ISO 14000	Not adopted	Low
Internal audits	Adopted: Power	High
External audits	Adopted: Power	High
Green products	Adopted: leadership	Low
Lilly		
Underground tanks	Adopted: leadership	Low
EPA's 33/50	Adopted: leadership	Low
Responsible Care	Initially not adopted	High
	Finally adopted: leadership	Low
ISO 14000	Not adopted	Low
Internal audits	Adopted: leadership	Low
External audits	Not adopted	High
Green products	n.a.	n.a.

Summary

Outcome/Process	Organizational changes are significant	Organizational changes are not significant
Policy is not adopted	External audits (Lilly) Responsible Care (Lilly)	ISO 14000 (Baxter, Lilly)
Policy is adopted by power-based processes	Internal and external audits (Baxter)	
Policy is adopted by leadership-based processes	Internal audits (Lilly) Responsible Care (Lilly)	Underground tanks (Baxter, Lilly) 33/50 (Baxter, Lilly) Green products (Baxter)

significant in two of them (external audits and phase one of Responsible Care). Consequently, these policies were not adopted. Organizational changes were less significant for the other four policies, three of which were adopted by leadership-based processes (underground Tanks, 33/50, and the second phase of Responsible Care), and one which was not adopted (ISO 14000).

Whether or not institutional changes are significant requires examining the nature of the institutions. The Institutional Analysis and Development (IAD) framework identifies three nested levels of institutional analysis: operational choice, collective choice, and constitutional choice (Kiser and Ostrom 1982). Constitutional choice level changes are the most difficult to implement while operational choice level changes are the easiest. The former alter the "status quo" most significantly since all future collective choice decisions would be affected by such changes. Consequently, they create incentives for "losers" to tenaciously resist them. In the early 1990s, Baxter established environmental auditing systems and invited Arthur D. Little (ADL) to define state-of-the-art standards. This action deprived Baxter's facility managers of a crucial voice in defining environmental standards. As a result, they opposed the concept of state-of-the-art standards and the invitation to ADL to define them. Such collective choice level changes are adopted only if their supporters are hierarchical superiors having the abilities to ignore opposition. Baxter's policy had vocal and decisive support from Vernon Loucks, Baxter's Chief Executive Officer, and Marshall Abbey, Chairman of Baxter's Environmental Review Board who virtually mandated the adoption of this policy.

In contrast, Lilly's policy on internal audits did not ignite internal opposition since it did not upset the status quo. Unlike Baxter, Lilly defined its auditing standards internally with contributions from both corporate and facility managers. In the early 1990s, Lilly employed environmental audits primarily as a compliance tool, and only in recent years have they been employed to evaluate management systems as well. Such changes in management systems create significant conflicts only if they alter organizational structures or reallocate functional responsibilities. In contrast to Lilly, Baxter did not adopt such an incremental policy on audits; rather, the compliance and management systems aspects were emphasized from the very beginning.

Operational choice level decisions do not significantly impact inter-manager interactions since they do not initiate new policies and they only marginally reallocate responsibilities among existing managers. In effect, they do not create significant losers. This is an important reason why the underground storage tank program and the 33/50 program in Baxter and Lilly, and the marketing of green products in Baxter, met with little resolute opposition. These programs did not require creating new permanent structures (although implementation committees were formed in both these firms) and therefore did not redistribute responsibilities or budgets within these firms. With managers not feeling professionally threatened, they had few incentives to tenaciously oppose the new policies.

On the other hand, some of Lilly's managers initially opposed Responsible Care, especially its Code on Community Awareness and Emergency Response. They questioned its usefulness and legitimacy. This Code required CMA members to develop Community Outreach Programs. The policy skeptics argued that Lilly had little reason to share information on its manufacturing operation since it adhered to all applicable laws and regulations. Further, communities in the vicinity of Lilly's facilities had little knowledge with which to appreciate the technicalities of manufacturing processes. A sharing of technical information with such communities may lead them to make unwarranted conclusions about the safety aspects and environmental impacts of facilities' operations. Opposition to Responsible Care stemmed from the requirement for Lilly to change its policy (collective choice level institution) on outsiders' access to information on Lilly's internal operations. Responsible Care required significant changes in a major policy and the affected internal managers had little voice in shaping this change. Policy skeptics viewed the code as fundamentally redefining the relationship between their firm and communities living in the vicinity of its facilities, and felt threatened by this redefinition.

This discussion has important implications for policy design: policy-makers should consider the impact of a proposed policy on organizational politics and the incentives for managers to support or oppose it. Policies that seek to draw upon firms' extant structures and institutions have greater likelihood of adoption and successful implementation. In contrast, if Type 2 policies require significant changes at the constitutional choice and collective choice levels, they will probably encounter stiff internal opposition. And perhaps, such opposition may only be overcome by employing power-based processes.

Limitations and future research

This study employed a new-institutionalist perspective to examine environmental policymaking in Baxter and Lilly. It integrated insights from sociological institutional theory and stakeholder theory with power-based and leadership-based theories. Intra-firm policy processes and inter-manager interactions can also be examined by employing other theories and frameworks such as resource-based (Penrose 1959; Peteraf 1993), evolutionary (Nelson and Winter 1982), garbage can (Cohen, March, and Olsen 1972), and ecology (Hannan and Freeman 1977). These theories do not employ the same set of assumptions as new-institutionalism. In particular, since most of them do not subscribe to methodological individualism, they do not treat institutions (policies of firms) as conscious artifacts that can be traced to the preferences, endowments, and strategies of individual actors. These theories could therefore suggest different explanations for why firms selectively adopt Type 2 policies. Hence, this study should be viewed as a modest attempt to examine the phenomenon of beyond-compliance policies by employing one of many possible perspectives.

I focused on Type 2 policies since they appear anomalous to neoclassical economic theory. My conclusions are not generalizable to all beyond-compliance policies that include Type 1 as well. On this count, this book should not be interpreted as a general critique of the neoclassical theory.

Transaction cost theories are an integral part of the new institutionalist framework. Coase (1937) and Williamson (1975, 1985) argue that firms and other institutions arise to minimize transaction costs associated with market exchanges. Transaction cost theory is therefore employed to understand structural changes such as evolution of "hierarchies." In general, many argue that institutions are efficiency-enhancing artifacts and they evolve since actors wish to reduce the transaction costs associated with collective action (for a review, see Eggertsson 1990). Future research could also focus on applying transaction cost theories to understand evolution of micro-firm-specific policies such as Type 2 policies. For example, it could be hypothesized that firms adopt those Type 2 policies that reduce the sum of transaction cost and transformation cost. Since the standard capital budgeting techniques cannot capture the impact of policies on transaction costs in the future, some ways to measure transaction costs are required to test such hypothesis.

Previously, I discussed the need to demythologize the adoption of beyond-compliance policies which should be understood only in the context of extant laws and regulations and the incentives for preempting more stringent laws or shaping future laws. Adopting Type 2 policies is a self-interested response by managers and their firms. Beyond-compliance policies are adopted in other areas as well such as workplace safety and consumer safety. It would therefore be instructive to compare the processes of beyond-compliance policymaking across issue areas within and across firms and examine the validity of my analysis. Further, I have not "unpacked" regulatory agencies (such as the EPA) and implicitly treated them as monoliths. An area of further research could be to examine how differences in regulatory styles within a given governmental agency impact dynamics within and across firms on beyond-compliance policies.

This study focused on policy adoption only. It has not examined the durability and efficacy of power-based policies versus leadership-based policies. Arguably, power-based policies are effective in the short-run but not in the long-run. Only a consensual route of policy adoption succeeds in the long-run given that skeptics often have ways to oppose, and if driven to desperation, even to sabotage it. On the other hand, it could be argued that if top management's intent is clearly communicated, and power-based policies are able to meet their stated objective, the skeptics may become reconciled to new rules of the game. There are theoretical reasons to argue in favor as well as against power-based policies in the context of their long-run durability and efficacy. However, this is an important issue worthy of more research.

In examining inter-manager interactions, the book classified managers as policy supporters and policy skeptics. Somewhat in the tradition of Samuelson's (1947) 'revealed preference,' it inferred managerial preferences for Type 2 policies from their behaviors. Methodologically, however, this is an imperfect way of assessing preferences since preferences and exhibited behaviors may not have direct correspondence. In other words, the same set of preferences can translate into different behaviors given varying incentives. In some instances, actors may not have opportunities to express their preferences or may choose not to do so due to fears of retribution.

Being conscious of such pitfalls, I examined the reasons why managers support or oppose a given policy. During my interviews, I gathered information on the professional backgrounds of managers working on environmental issues across functional areas. From these interviews it was clear that most of them had strong personal commitments to environmental issues. This was not surprising since they chose to work on environmental issues; not because of organizational-level incentives such as special allowances or faster promotions. Many of these managers also claimed to be "environmentalists": some of them claimed to support the Sierra Club and other environmental groups financially. I also observed posters on environmental issues adorning offices of most of these managers. Most of them are very active in Earth Day celebrations. Such a display of support for environmental issues and my numerous discussions with them leads me to infer that most of these managers indeed hold strong beliefs on environmental issues. Thus, I expect that such managers would support Type 2 policies. On the other hand, I also expect "losers" from any organizational change that results from a Type 2 policy to assume the roles of skeptics. These losers could belong to environmental departments or work on environmental issues in other departments. Nevertheless, the strategy of inferring preferences from behaviors is not methodologically perfect.

This book argued that Type 2 policies are adopted only through power-based or leadership-based processes. It could be argued that leadership-based processes are in fact power-based processes where power is exercised subtly in terms of shaping opinion. As discussed in chapter 2, scholars following the perspectives of Gramsci (1988) and Foucault (1970) could also be expected to argue along the same lines. For example, it could be argued that efforts of policy supporters such as having their firm's Washington lobbyists make presentations on the desirability of adopting Type 2 policies (as in underground tanks) are really strategies to manipulate opinions of the skeptics. In this context, power is being subtly exercised.

If power is defined in such an all-encompassing manner, it becomes impossible to make falsifiable predictions about whether or not power has been exercised. The book defined the outcomes of power-based processes in a specific way: preferences and exhibited behaviors of policy skeptics do not change. If the ability to influence preferences (as in leadership-based processes) represents an exercise of power, then leadership is indeed a manifestation of power. I, however, have chosen not to define power in this manner.

Power-based and leadership-based processes are indeed different for the following reasons. First, in power-based processes the evidence of dissent is very clear. Such continued opposition of policy skeptics became even more evident during interviews. Once I had won their confidence, especially in terms of protecting confidentiality, I came across a wealth of information on their continued negative perceptions about Type 2 policies. It then became easier for me to separate the instances where preferences had changed due to induced cooperation from where preferences had not changed at all. Further, in both these firms, most managers are fairly independent and outspoken; if they disagree, they speak out. Such public displays of disagreements have become even more pronounced since both of these firms are attempting to decentralize and to empower their division- and facility-level managers.

The book suggests that power-based and leadership-based explanations *together* explain why firms selectively adopt Type 2 policies. It could, however, be argued that senior managers first employ a leadership-route, and if it does not work out, resort to power-based processes. Hence, power-based and leadership-based processes are sequential. I disagree with this as well. When policy supporters in Baxter decided to invite external auditors to define and then to audit facilities against the state-of-the-art standards, they adopted a power-based route: Vernon Loucks and Marshall Abbey mandated its implementation. Why was a leadership-based route not adopted? The reason is that since the policy supporters had committed a very tight implementation schedule to the Environmental Review Board, they did not want to experiment with a potentially time-consuming consensual route. As a result, they opted for employing the power of top management fiat. This suggests that the strategy employed by policy supporters is not sequential; it depends on a variety of factors such as the time table for implementation, the level of perceived opposition, and the perceived success of employing fiat.

This book operationalized procedural efficiency in terms of the most well-accepted and widely used procedure of investment analysis – capital budgeting – employed by large US firms. To make *ex ante* falsifiable predictions that firms are efficient institutions, efficiency needs to be defined in a precise manner. Otherwise, most policies would be classified *ex post* as being efficient, robbing the concept of efficiency of its analytical power. Other investment procedures such as full-cost accounting and life-cycle analysis lack acceptability among managers, were not employed by either Lilly or Baxter, and, importantly, do not lead to the maximization of shareholders' wealth, the avowed objective in the neoclassical theory. The book, therefore, operationalized procedural efficiency in terms of capital budgeting. This procedure seeks to maximize shareholder's wealth, an important objective geared for minimizing principal–agent conflicts in large corporations where principals (shareholders) do not have a direct say in the running of the firm and are vulnerable to agent (managers) abuse. However, in owner-managed businesses where such principal-agent conflicts are not significant, capital budgeting may not be employed. It is also difficult to identify other procedures that are consistently employed by owner-managers to *ex ante* estimate profitability of projects. Nevertheless, this does not imply that such businesses are inefficient. Thus, the findings and conclusions of this book are applicable only to large corporations that employ capital budgeting as an established tool for investment analysis.

In spite of their widespread acceptability in large US corporations, discounted cash flow techniques have been widely criticized. Some scholars argue that an over-reliance on them has lead to decline of America's competitive position (Donaldson 1972; Hayes and Abernathy 1980; Hout, Porter, and Rudden 1982). They point out methodological biases in capital budgeting that distort investment decisions, thereby forcing managers to adopt short-term horizons (Hodder and Riggs 1985; Kulatilaka and Marcus 1992). Clearly, with different national systems of industrial organization, capital budgeting is perhaps not a widely accepted tool in say the Japanese *Keiretsus* (Hodder 1984). Most Japanese firms are over-capitalized by US standards, an outcome that would be discouraged by capital budgeting procedures. It is also suggested that Japanese managers have long time horizons, while capital budgeting encourages shorter time horizons. Further, since Japanese managers are perceived as being relatively insulated from the pressures of the stock markets, it is suggested that they would not be unduly worried about ensuring adequate rates of return on investment. Kester notes that:

Japanese corporate governance emphasizes the reduction in transaction costs associated with self-interested opportunism and investment in relations-specific asset. This strategy fosters the building of stable, long-term commercial relationships among transacting companies, *although general (i.e., noncorporate, nonlending) shareholders are often forced to bear potentially substantial agency costs.* Anglo-American corporate governance, in contrast, emphasizes the reduction in agency costs associated with the separation of ownership from control, relying more heavily on formal, legalistic mechanisms to order commercial relationships among transacting parties (1996: 108; italics mine).

Hence, capital budgeting may be inadequate, if not the wrong route, to operationalize efficiency in the Japanese or other non-US systems of industrial organizations. On this count, my findings and conclusions may lack generalizability to non-US contexts.

As discussed in chapter 1, the 1993 UNCTAD benchmark survey suggests that environmental programs of MNEs are critically influenced by three factors: sales volume, nationality of their parent firm, and industry type. Baxter and Lilly share these characteristics in that both have 1996 sales greater than $4.9 billion, both have American parents, and both are in similar industries. On this count, the implication of this research for theory development and policymaking may not be generalizable to firms that vary on one or more of these three attributes. Consequently, future research could investigate the implications of this research by studying firms that vary on these three attributes.

References

Ackerman, R. W. 1975. *The Social Challenge to Business.* Cambridge, MA: Harvard University Press.

Alchian, A. A. 1950. Uncertainty, evolution, and economic theory. *Journal of Political Economy,* 58: 211–21.

Allegiance. 1997. http://www.allegiance.net/COMPNEWS/Feb0397.HTM; page 2 of 3; 04/27/97.

Allison, G. T. 1971. *Essence of Decision: Explaining the Cuban Missile Crisis.* Boston, MA: Little Brown.

Ambrose, S. E. 1967. *Eisenhower and Berlin.* New York: W.W. Norton.

Appleby, J. O., Hunt, L., and Jacob, M. 1994. *Telling the Truth About History.* New York: Norton.

Arkansas Gazette. 1992. March 19: Arkansas Page.

Arora, S. and Cason, T. N. 1996. Why do firms volunteer to exceed environmental regulations? Understanding participation in EPA's 33/50 program. *Land Economics,* 72: 413–32.

Arrow, K. J. 1987. Reflections on the essays. In G. Feiwel (ed.), *Arrow and the Foundations of the Theory of Economic Policy.* New York: New York University Press, pp. 727–34.

Arthur D. Little, Inc. 1995. Hitting the green wall. *Perspectives.*

Atkinson, S. and Tietenberg, T. H. 1991. Market failures in incentive-based regulations: The case of emission trading. *Journal of Environmental Economics,* 21: 17–31.

Barnard, C. I. 1938. *The Functions of an Executive.* Cambridge, MA: Harvard University Press.

Barrett, S. 1991. Environmental regulations for competitive advantage. *Business Strategy Review,* 2: 1–15.

Barton, S. L., Hill, N. C., and Sundaram, S. 1989. An empirical test of stakeholder theory predictions of capital structure. *Financial Management,* 18: 36–44.

Bass, B. M. 1985. *Leadership and Performance Beyond Expectations.* New York: Free Press.

Bates, R. H. 1983. *Essays on the Political Economy of Rural Africa.* Cambridge University Press.

Baumol, W. J. 1959. *Business Behavior, Value, and Growth.* New York: MacMillan.
1982. *Economic Theory and Operations Analysis.* New Delhi: Prentice Hall.

Baumol, W. J. and Oates, W. 1988. *The Theory of Environmental Policy.* Cambridge University Press.

Baxter Environmental Manual/Baxter. 1994. Deerfield, IL: Baxter Healthcare Corporation

Baxter Healthcare Corporation/Baxter. 1985a. *Travenol Environmental Newsletter.* October. Deerfield, IL: Baxter Healthcare Corporation

1985b. *Travenol Environmental Newsletter.* December. Deerfield, IL: Baxter Healthcare Corporation

1986 *Travenol Environmental Newsletter.* October. Deerfield, IL: Baxter Healthcare Corporation

1988. *Baxter Environmental Review.* September. Deerfield, IL: Baxter Healthcare Corporation

1989. *Baxter Environmental Review.* October. Deerfield, IL: Baxter Healthcare Corporation

1990a. *Baxter Environmental Review.* May. Deerfield, IL: Baxter Healthcare Corporation

1990b. *Environmental Leadership: Remarks by Vernon R. Loucks Jr., Chairman and Chief Executive Officer, Baxter International.* Deerfield, IL: Baxter International, Inc.

1991a. *Baxter Environmental Review.* May. Deerfield, IL: Baxter Healthcare Corporation

1991b. *Baxter Environmental Review.* September. Deerfield, IL: Baxter Healthcare Corporation

1992a. *Serving the Environment: A Call to Action for the Health Care Industry.* Deerfield, IL: Baxter International, Inc.

1992b. *Baxter Environmental Review.* May. Deerfield, IL: Baxter Healthcare Corporation

1992c. *Baxter Environmental Review.* August. Deerfield, IL: Baxter Healthcare Corporation

1993. *Environmental Highlights.* Deerfield, IL: Baxter Healthcare Corporation

1994a. *Annual Report.* Deerfield, IL: Baxter International, Inc.

1994b. *Environmental Performance Report.* Deerfield, IL: Baxter International, Inc.

1994c. *Baxter Environmental,* 2(3). Deerfield, IL: Baxter International, Inc.

1994d. *Baxter Environmental Initiatives for Customers.* Deerfield, IL: Baxter International, Inc.

1995a. *State-of-the-art Environmental Program Report 1994.* Deerfield, IL: Baxter International, Inc.

1995b. *Baxter Environmental,* 3(2). Deerfield, IL: Baxter International, Inc.

1996a. *Environmental Performance Report 1995.* Deerfield, IL: Baxter International, Inc.

1996b. *Internal Memo on "Customer Environmental Initiatives."* May 16. Deerfield, IL: Baxter International, Inc.

1996c. *Annual Report.* Deerfield, IL: Baxter International, Inc.

1997a. http://www.baxter.com/www/financia...port/1996/financialhighlights. html; page 1 of 2; 04/27/97.

1997b. *Environmental, Health and Safety Performance Report.* Deerfield, IL, Baxter International, Inc.

Berle, A.A. and Means, G.C. 1932. *The Modern Corporations and Private Property.* New York: MacMillan.

Bernstein, M. H. 1955. *Regulating Business by Independent Commission*. Princeton, NJ: Princeton University Press.

Bessen, S. M. and Saloner, G. 1988. *Compatibility Standards and the Market for Telecommunication Services*. Santa Monica, CA: Rand.

Boulding, K. E. 1963. Towards a pure theory of threat systems. *American Economic Review*, 53: 424–34.

Bowles, S. and Gintis, S. 1993. The revenge of homo economicus: Contested exchange and the revival of political economy. *Journal of Economic Perspectives*, 7: 83–102.

Buchanan, J.M. 1962. Marginal notes on reading political philosophy. In J. M. Buchanan and G. Tullock. *A Calculus of Consent*. Ann Arbor, MI: University of Michigan Press, pp.307–22.

  1965. An economic theory of clubs. *Economica*, 32: 1–14.

Bunge, J., Cohen-Rosenthal, E., and Ruiz-Quintanilla, A. 1995. Employee participation in pollution reduction. New York State School of Industrial and Labor Relations and Cornell University, Mimeo.

Carroll, A. B. 1995. Stakeholder thinking in three models of management morality: A perspective with strategic implications. Reprinted in M.B.E. Clarkson, (ed.), *The Corporation and its Stakeholders*. Toronto: University of Toronto Press, pp. 139–70.

Carson, R, 1962, *Silent Spring*. Greenwich, CO: Fawcett.

Cascio, J. 1994. International environmental management standards – ISO 9000s less tractable siblings. *ASTM Standardization News*, April: 44–9.

Chandler, A. D. 1962. *Strategy and Structure*. Cambridge, MA: MIT Press.

  1977. *The Visible Hand*. Cambridge, MA: Harvard University Press.

  1980. Government versus business: An American phenomenon. In J. T. Dunlop (ed.) *Business and Public Policy*. Cambridge, MA: Harvard University Press, pp. 1–11.

Charter, M. (ed.). 1992. *Greener Marketing*. Sheffield: Greenleaf Publishing.

Chatterjee, P. and Finger, M. 1994. *The Earth Brokers: Power, Politics, and Development*. London and New York: Routledge.

Chemers, M.M. and Ayman, R. (eds.) 1993. *Leadership Theory and Research: Perspectives and Directions*. San Diego, CA: Academy Press.

Chemical and Engineering News. 1992. Chemical makers pin hope on Responsible Care to improve image. October 5: 13–39.

  1993a. Responsible Care program poses challenges for smaller firms. August 9: 9–14.

  1993b. New initiatives take aim at safety performance of chemical industry. November 29: 12–36.

Chemical Manufacturers Association/CMA. 1996a. *About CMA*. http://www.cmahq.com/abtcma.html, 12/27/96.

  1996b. *News and Issues*. http://www.cmahq.com/news1.html, 12/27/96.

  1998. *Management System Verification*. http://www.cmahq.com/cmapgrams; 1/16/98.

  1999. *Community Right to Know*. http://www.cmahq.com/issueadvocay.nsf; 03/12/99.

Chemical Week. 1995. Managers gear up for global standards. October 11: 65–6.

Child, J. 1972. Organizational structure, environment, and performance – The role of strategic choice. *Sociology*, 6: 1–22.

Choucri, N. (ed.) 1993. *Global Accord: Environmental Challenges and International Response*. Cambridge, MA: MIT Press.

Clarkson, M.B.E. 1995. A stakeholder framework for analyzing and evaluating corporate social performance. *Academy of Management Review*, 20: 92–117.

Coase, R. H. 1937. The nature of the firm. *Economica*, 4: 386–405.

 1988. *The Firm, the Market, and the Law*. Chicago, IL: University of Chicago Press.

 1993.The institutional structure of production: 1991 Noble Prize lecture. In O. E. Williamson and S. Winter (eds.), *The Nature of the Firm*. New York: Oxford University Press, pp. 227–35.

Cochran, P.L. and Wood, R. A. 1984. Corporate social responsibility and financial performance. *Academy of Management Journal*, 27: 42–56.

Coddington, W. 1993. *Environmental Marketing: Positive Strategies for Reaching the Green Consumer*. McGraw-Hill.

Cohen, M.D., March, J.G., and Olsen, J.P. 1972. A garbage can model of organizational choice. *Administrative Science Quarterly*, 17: 1–25.

Colborn, T. 1996. *Our Stolen Future: Are We Threatening Our Fertility, Intelligence, and Survival: A Scientific Detective Story*. New York: Dutton.

Commons, J. R. 1968. *Legal Foundations of Capitalism*. Madison, WI: University of Wisconsin Press.

Cornes, R. and Sandler, T. 1996. *The Theory of Externalities, Public Goods, and Club Goods*. Cambridge University Press.

Crawford, S. E. S. and Ostrom, E. 1995. A grammar of institutions. *American Political Science Review*, 89: 582–600.

Cross, J.G. and Guyer, M. J. 1980. *Social Traps*. Ann Arbor, MI: University of Michigan Press.

Currie, R. M. 1995. 33/50 program progress update, December 1995. A note prepared for submission to the EPA, December 20, 1995. Deerfield, IL: Baxter Healthcare Corporation.

Cyert, R. M. and March, J. G. 1963. *A Behavioral Theory of Firm*. Englewood Cliffs, NJ: Prentice Hall.

Dahl, R. A. 1957. The concept of power. *Behavioral Science*, 2: 201–15.

Davis, T. R. V. and Luthans, F. 1980. A social learning approach to organization behavior. *Academy of Management Review*, April: 281–90.

Deming, W. E. 1982. *Quality, Productivity, and Competitive Position*. Cambridge, MA: Massachusetts Institute of Technology, Center for Advanced Engineering Study.

 1992. *The World of W. Edward Deming*. Compiled by C. S. Kilian. Knoxville, TN: SPS Press.

Demsetz, H. and Lehn, K. 1985. The structure of corporate ownership: Causes and consequences. *Journal of Political Economy*, 93: 1155–77.

DiMaggio, P. and Powell, W. W. (eds.) 1983. *The New Institutionalism in Organizational Analysis*. Chicago, IL: University of Chicago Press.

Dodd, L. and Schot, R. 1979. *Congress and the Administrative State*. New York: Wiley.

Donaldson, G. 1972. Strategic hurdle rates for capital investment. *Harvard Business Review*, 50: 50–5.

Donaldson, T. and Preston, L.E. 1995. The stakeholder theory of the corporation: Concepts, evidence, and implications. *Academy of Management Review*, 20: 65–91.

Dow, G. 1987. The function of authority in transaction costs economics. *Journal of Economic Behavior and Organization*, 8: 13–38.

Dugger, W. 1983. The transaction cost analysis of Oliver Williamson: A new synthesis? *Journal of Economic Issues*, 17: 95–114.

Dunning, J. 1993. *The Globalization of Business*. London: Routledge.

Edwards, R. 1979. *Contested Terrain: The Transformation of the Workplace in the Twentieth Century*. New York: Basic Books.

Eggertsson, T. 1990. *Economic Behavior and Institutions*. Cambridge University Press.

1996. The old theory of economic policy and New Institutionalism. Presented at the Workshop on Economic Transformation, Institutional Change, Property Rights, and Corruption, March 7–8, 1996, Washington, DC.

Environmental Protection Agency/EPA. 1994. *Questions and answers about the 33/50 program*. http://es.inel.gov/partners/3350/q-a.html, 12/11/96.

Enzi, M. 1997. EPA's obstruction of pollution control. *Wall Street Journal*, April 13: A18.

Esty, D. C. 1994. *Greening the GATT*. Washington, DC: Institute for International Economics.

Etzioni, A. 1988. *The Moral Dimension*. New York: Basic Books.

Fama, E. 1980. Agency problem and the theory of the firm. *Journal of Political Economy*, 88: 288–307.

Fiedler, F. E. 1967. *A Theory of Leadership Effectiveness*. New York: McGraw Hill.

Fischer, K. and Schot, J. (eds.) 1992. *Environmental Strategies for Industries*. Washington, DC: Island Press.

Follett, M. P. 1940. Constructive conflict. In H.C. Metcalf and L. Urwick (eds.), *Dynamic Administration*. New York: Harper and Row Publishers, pp. 30–49.

Foucault, M. 1970. *The Order of Things*. Translated from French. London: Tavistock Publishers.

Freeman, R.E. 1984. *Strategic Management: A Stakeholder Approach*. Boston: Pittman.

Fri, R. W. 1992. The corporation as a nongovernmental organization. *The Columbia Journal of World Business*, 27: 91–5.

Friedman, M. 1970. The social responsibility of business is to increase its profits. *New York Times Magazine*, September 13: 32–3, 122–6.

Furubotn, E. G. and Richter, R. 1991. The new institutional economics: An assessment. In E. G. Furubotn and R. Richter (eds.), *The New Institutional Economics*. College Station: Texas AandM University Press, pp. 1–32.

Gabel, H. L. 1994. From market failure to organization failure. *Business Strategy and the Environment*, 3.

Georgopolous, B. S., Mahoney, G.M. and Jones, N. W. 1957. A path-goal approach to productivity. *Journal of Applied Psychology*, December: 345–55.

Ghoshal, S. and Moran, P. 1996. Bad for practice: A critique of the transaction cost theory. *The Academy of Management Review*, 21: 13–47.

Graen, G., Novak M., and Sommerkamp, P. 1982. The effects of leader-member exchange and job design and productivity and satisfaction: Testing a dual attachment model. *Organizational Behavior and Human Performance,* 30: 10–131.

Gramsci, A. 1988. *An Antonio Gramsci Reader.* D. Forgacs (ed.) New York: Schocker Books.

Granovetter, M. 1985. Economic action and social structure: The problem of embeddedness. *American Journal of Sociology,* 91: 481–510.

Greene, C. N. 1975. The reciprocal nature of influence between leader and subordinate. *Journal of Applied Psychology,* 60: 187–93.

Griffin, J. J. and Mahon, J.F. 1997. The corporate social performance and corporate financial performance debate: Twenty-five years of incomparable research. *Business and Society,* 36: 5–31.

Hahn, R. and Nell, R. 1982. Designing a market for tradable emission permits. In W. Megat (ed.), *Reform of Environmental Regulations.* Cambridge, MA: Ballinger.

Hajer, M.A. 1995. *The Politics of Environmental Discourse.* Oxford: Clarendon Press.

Hall, P. A. (ed.) 1986. *The Political Power of Economic Ideas.* Princeton, NJ: Princeton University Press.

Hannan, M. T. and Freeman, J.H. 1977. The population ecology of organization. *American Journal of Sociology,* 82: 929–64.

Hardin, G. 1968. The tragedy of commons. *Science,* 162: 1243–48.

Hart, S. 1995. A natural resource-based view of the firm. *Academy of Management Review,* 20: 986–1014.

Hayek, F. A. 1945. The use of knowledge in society. *American Economic Review,* 35: 519–39.

 1955. *The Counterrevolution of Science.* New York: Free Press.

Hayes, R. and Abernathy, W. 1980. Managing our way to economic decline. *Harvard Business Review,* 58: 67–77.

Herzberg, F. 1966. *Work and the Nature of Man.* Cleveland: World Publishing Company.

Hirshleifer, J. 1988. *Price Theory and Its Applications.* 4th edn., Englewood Cliffs: Prentice Hall.

Hodder, J. E. and Riggs, H. E. 1985. Pitfalls in evaluating risky projects. *Harvard Business Review,* 85: 128–35.

Hodder, J. F. 1984. Evaluation of manufacturing investments: A comparison of US and Japanese practices. *Technical Report 84–8,* Department of Industrial Engineering and Engineering Management, Stanford University.

Hoffman, A. J. 1997. *From Heresy to Dogma.* San Francisco: New Lexington Press.

House, R. J. 1971. A path-goal theory of leader effectiveness. *Administrative Science Quarterly,* September: 321–38.

 1976. A theory of charismatic leadership. In J. G. Hunt and L. L. Larson (eds.), *Leadership: The Cutting Edge.* Carbondale, IL: Southern Illinois University Press, pp. 189–207.

Hout, T., Porter, M. and Rudden, E. 1982. How global companies win out. *Harvard Business Review,* 60: 98–108.

Ikenberry, G. J. 1993. Creating yesterday's new world order. In J. Goldstein and R.

O. Keohane (eds.), *Ideas and Foreign Policy*. Ithaca, NY: Cornell University Press, pp. 57–86.

International Organization for Standardization. 1995a. *ISO/DIS 14001: Environmental management systems–general guidelines on principles, systems, and supporting techniques*. Draft. Geneva: International Organization for Standardization.

1995b. *ISO/DIS 14004: Environmental management systems–general guidelines on principles, systems, and supporting techniques*. Draft. Geneva: International Organization for Standardization.

Jaffe, B. A., Peterson, S. R., Portney, P. R., and Stavins, R. N. 1995. Environmental regulation and the competitiveness of US manufacturing: What does the evidence tell us? *Journal of Economic Literature*, 33: 132–63.

Jensen, M. 1983. Organization theory and methodology. *Accounting Review*, 58: 319–39.

Jensen, M. and Meckling, W. 1976. Theory of firm: Managerial behavior, agency cost, and ownership structure. *Journal of Financial Economics*, 3: 305–60.

Johnson, P. L. 1994. *ISO 9000: Meeting the New International Standards*. McGraw-Hill.

Jones, T.M. 1995. Instrumental stakeholder theory: A synthesis of ethics and economics. *Academy of Management Review*, 20: 404–37.

Katz, D. and Kahn, R. L. 1966. *The Social Psychology of Organizations*. New York: Wiley.

Katz, M. L. and Shapiro, C. 1983. *Network Externalities, Competition, and Compatibility*. Princeton, NJ: Woodrow Wilson School, Princeton University.

Katz, M. L. and Shapiro, C. 1985. *Technology Adoption in the Presence of Network Externalities*. Princeton, NJ: Woodrow Wilson School, Princeton University.

Katz, R. 1974. Skills of an effective administrator. *Harvard Business Review*, September– October: 90–101.

Katzmann, R. 1980. *Regulatory Bureaucracy: The Federal Trade Commission and the Antitrust Policy*. Cambridge, MA: MIT Press.

Kulatilaka, N. and Marcus, A. J. 1992. Project valuation under uncertainty: When does DCF fail? *Journal of Applied Corporate Finance* 5: 92–100.

Keohane, R. O. 1988. International institutions: Two approaches. *International Studies Quarterly*, 32: 379–96.

Kester, W. C. 1996. American and Japanese corporate governance: Convergence to best practice. In S. Berger and R. Dore (eds.), *National Diversity and Global Capitalism*. Ithaca, NY: Cornell University Press, pp.107–37.

Khademian, A. M. 1992. *The SEC and Capital Market Regulation: The Politics of Expertise*. Pittsburgh: University of Pittsburgh Press.

Khanna, M. and Damon, L.A. 1999. EPA's voluntary 33/50 program: Impact on toxic releases and economic performances of firms. *Journal of Environmental Economics and Management*, 37: 1–25.

King, G., Keohane, R. O., and Verba, S. 1994. *Designing Social Inquiry*. Princeton, NJ: Princeton University Press.

Kiser L. L. and Ostrom, E. 1982. Three worlds of collective action: A metatheoretical synthesis of institutional approaches. In V. Ostrom (ed.), *Strategies for Political Inquiry*. Beverly Hills, CA: Sage, pp. 179–222.

Kissinger, H. A. 1964. *A World Restored.* New York: Grosset and Dunlap.

Knight, J. 1992. *Institutions and Social Conflict.* Cambridge University Press.

Kolko, G. 1963. *Railroads and Regulation, 1877–1916.* Princeton, NJ: Princeton University Press.

Kotter, J. and Heskett, J. 1992. *Corporate Culture and Performance.* New York: Free Press.

Kreps, D. M. 1990. Corporate culture and economic theory. In J. Alt and K. Shepsle (eds.), *Perspectives on Positive Political Economy.* Cambridge University Press, pp. 90–143.

Lee, D. and Misiolek, W.S. 1986. Substituting pollution taxation for general taxation: Some implications for efficiency in pollution taxation. *Journal of Environmental Economics and Management,* 13: 338–47.

Leibenstein, H. 1966. Allocative efficiency and X efficiency. *The American Economic Review,* 56: 392–415.

Leonard, J. H. 1988. *Pollution and the Struggle for the World Product: Multinational Corporations, the Environment, and International Comparative Advantage.* Cambridge University Press.

Libecap, G. D. 1989. *Contracting for Property Rights.* Cambridge University Press.

Lilly. 1992a. *A Perspective on Pricing.* Indianapolis, IN: Eli Lilly and Company.

1992b. *Environmental Annual Report.* Indianapolis, IN: Eli Lilly and Company.

1993. *Environmental Annual Report.* Indianapolis, IN: Eli Lilly and Company.

1994a. *Annual Report.* Indianapolis, IN: Eli Lilly and Company.

1994b. *Environmental Annual Report.* Indianapolis, IN: Eli Lilly and Company.

1994c. *Responsible Care: A Public Commitment.* Indianapolis, IN: Eli Lilly and Company.

1995a. *Environmental, Health, and Safety Report.* Indianapolis, IN: Eli Lilly and Company.

1995b. *Information 1995.* Indianapolis, IN: Eli Lilly and Company.

1995c. *Environmental Annual Report.* Indianapolis, IN: Eli Lilly and Company.

1996a. *The Three Loop Operations Design Model for Environmental, Health, and Safety Programs.* Indianapolis, IN: Eli Lilly and Company.

1996b. *Environmental Policies and Guidelines.* Indianapolis, IN: Eli Lilly and Company.

1997a. http://www.lilly.com/financial/95-annual/html/hilites.html; 03/02/97.

1997b. http://www.lilly.com/environment/96update/96gr9_1.htm; 05/02/97.

Lipman-Blumen, J. 1996. *The Connective Edge: Leading in an Interdependent World.* Jossey-Bass.

Lowi, T. J. 1969. *The End of Liberalism: The Second Republic of the United States.* 2nd edn, New York: Norton.

Luthans, F. 1995. *Organizational Behavior.* Singapore: McGraw-Hill, Inc.

Majumdar, R. C., Raychaudhuri, H. C., and Datta, K. 1958. *An Advanced History of India.* 2nd edn, London: Macmillan.

Maloney, M. T. and McCormick, R. E. 1982. A positive theory of environmental quality regulation. *Journal of Law and Economics,* 25: 99–123.

Manne, H. 1965. Mergers and the market for corporate control. *Journal of Political Economy,* 73: 110–120.

March, J. and Simon, H. A. 1958. *Organizations*. New York: John Wiley.

March, J. G. and Olsen, J. P. 1989. *Rediscovering Institutions: The Organizational Basis of Politics*. New York: Free Press.

Marcus, A. A. 1984. *The Adversary Economy*. Westport, CO: Quorum Books.

Marglin, S. 1974. What bosses do? The origins and functions of hierarchy in capitalist production. *The Review of Radical Political Economy*, 6: 33–60.

Marris, R. 1964. *The Economic Theory of Managerial Capitalism*. London: MacMillan.

Maslow, A. H. 1943. A theory of human motivation. *Psychological Review*, July: 370–96.

Meyer, J. and Scott, W.R. (eds.) 1992. *Organizational Environments: Ritual and Rationality*. Thousand Oaks, CA: Sage.

Miller, G. J. 1992. *Managerial Dilemmas: Political Economy of Hierarchy*. Cambridge University Press.

Mitchell, R. K., Agle, B.R., and Wood, D.J. 1997. Towards a theory of stakeholder identification and salience: Defining the principle who and what really counts. *Academy of Management Review*, 22: 853–86.

Morandi, L. and Pascal, S. 1995. *Environmental Audits: Incentives to Comply with to Avoid Regulation?* February, Washington, DC : National Conference of State Legislatures.

Morgenthau, H. J. 1978. *Politics Among Nations: The Struggle for Power and Peace*. New York: Knopf.

Mueller, D. C. 1989. *Public Choice II*. Cambridge University Press.

National Academy of Public Administration (NAPA). 1986. *A Strategy for Implementing Federal Regulations of Underground Storage Tanks*. Washington, DC : National Academy of Public Administration.

Nehrt, C. 1998. Maintainability of first-mover advantages when environmental regulations differ between countries. *Academy of Management Review*, 23: 77–97.

Nelson, R. R. and Winter, S. G. 1982. *An Evolutionary Theory of Economic Change*. Cambridge, MA: Harvard University Press.

Newton, T. and Harte, G. 1996. Green business: Technicist kitsch? *Journal of Management Studies*, 34: 75–98.

New York Times. 1996a. Many states give polluting firms new protection. April 17: A1, A16.

1996b. December 26: A26.

North, D. C. 1990. *Institutions, Institutional Change, and Economic Performance*. Cambridge University Press.

Northouse, P.G. 1997. *Leadership: Theory and Practice*. Thousand Oaks, CA: Sage.

NSF International. 1996. *Environmental Management System Demonstration Project: Final Report*. Ann Arbor, MI: NSF International.

Oates, W. E., Portney, P. R., and McGartland, A. M.1989. The net benefits of environmental regulations. *American Economic Review*, 79: 1233–42.

Oblak, D. J. and Helm, R. J. 1980. Survey and analysis of capital budgeting methods used by multinationals. *Financial Management*, 2: 37–40.

Oliver, C. 1991. Strategic responses to institutional processes. *Academy of Management Review*, 16: 145–79.

Olson, M. 1965. *The Logic of Collective Action–Public Goods and the Theory of Groups.* Cambridge, MA: Harvard University Press.

Osborne, D. and Gaebler, T. 1992. *Reinventing Government: How the Entrepreneurial Spirit is Transforming the Public Sector.* Cambridge, MA: Addison-Wesley.

Ostrom, E. 1986. An agenda in the study of institutions. *Public Choice,* 48: 3–25.

1990. *Governing the Commons: Evolution of Institutions for Collective Action.* Cambridge University Press.

1991. Rational-choice theory and institutional analysis: Towards complementarity. *American Political Science Review,* 85: 237–43.

1996. Institutional rational choice: An assessment of the IAD framework. Working Paper W96-16, Bloomington, IN: Workshop in Political Theory and Policy Analysis.

Ostrom, E., Gardner, R., and Walker, J. 1994. *Rules, Games, and Common-Pool Resources.* Ann Arbor, MI: University of Michigan Press.

Ostrom, V. and Ostrom, E. 1977. Public goods and public choice. In E. S. Savas (ed.), *Alternatives for Delivering Public Services.* Boulder, CO: Westview, pp. 7–49.

Palmer K., Oates, W., and Portney, P. R. 1995. Tightening environmental standards: The benefit–cost or the no–cost paradigm? *Journal of Economic Perspectives,* 9: 119–32.

Pearson, C. S. 1985. *Down to Business: Multinational Corporations, the Environment, and Development.* Washington, DC: World Resource Institute.

Penrose, E. T. 1959. *The Theory of the Growth of the Firm.* London: Basil Blackwell.

Perrow, C. 1979. *Complex Organizations: A Critical Essay.* Glenview, IL: Scott, Foresman and Company.

Peteraf, M. 1993. The cornerstone of competitive advantage: A resource-based view. *Strategic Management Journal,* 14: 179–92.

Pfeffer, J. 1981. *Power in Organizations.* Marshfield, MA: Pitman.

Pfeffer, J. and Salanick. 1978. *The External Control of Organization.* New York: Harper and Row.

Pigou, A.C. 1960. *The Economics of Welfare.* London: Macmillan.

Platt, J. 1973. Social traps. *American Psychologist,* 26: 641–51.

Porter, M. E. 1991. America's green strategy. *Scientific American,* April: 168.

Porter, M. E. and van der Linde, C. 1995. Toward a new conception of the environment-competitiveness relationship. *Journal of Economic Perspectives,* 9: 97–118.

Prahalad, C.K. and Hamel, G. 1990. The core competence of the corporation. *Harvard Business Review,* 68: 79–91.

Prakash, A. 1999a. A new-institutionalist perspective on ISO 14000 and Responsible Care. *Business Strategy and the Environment.* Forthcoming.

1999b. Why firms adopt beyond-compliance policies. Mimeo.

Prakash, A. Krutilla, K., and Karamanos, P. 1996. Multinational corporations and international environmental policy. *Business and the Contemporary World,* 8: 119–44.

Preston, L.E. 1975. Corporation and society: The search for a paradigm. *Journal of Economic Literature,* 13: 434–53.

Preston, L.E. and Post, J.E. 1975. *Private Management and Public Policy: The Principle of Public Responsibility.* Englewood Cliffs, NJ: Prentice-Hall.

Preston, L.E. and Sapienza, H. J. 1990. Stakeholder management and corporate performance. *Journal of Behavioral Economics,* 19: 361–75.

Puri, S. C. 1996. *Stepping Up to ISO 14000: Integrating Environmental Quality with ISO 9000 and TQM.* Portland, OR: Productivity Press.

Putterman, L. 1984.On some recent explanations of why capital hires labor. *Economic Inquiry,* 22: 171–87.

Reitz, H. J. 1987. *Behavior in Organizations.* Homewood, IL.: Irwin.

Roberts, P. W. and Greenwood, R. 1997. Integrating transaction costs and institutional theories. *Academy of Management Review,* 22: 346–73.

Ross, S.A. 1974. The economic theory of agency: The principal problem. *American Economic Review,* 62: 134–9.

Rugman, A.R. and Verbeke, A. 2000. Environmental regulations and global strategies of multinational corporations. In A. Prakash and J. A. Hart(eds.), *Coping with Globalization.* London and New York: Routledge.

Russo, M.V. and Fouts, P.A. 1997. A resource-based perspective on corporate environmental performance and profitability. *Academy of Management Journal,* 40: 534–59

Saloner, Garth and Shepard, Andrea. 1991. *Adoption of Technologies with Network Effects: An Empirical Examination of the Adoption of Automated Teller Machines,* Stanford, CA: Graduate School of Business.

Salop, S.C. and Scheffman, D.T. 1983. Raising rivals' costs. *American Economic Association, Papers and Proceedings,* 73: 267–71.

Samuelson, P. A. 1947. *Foundations of Economic Analysis.* Cambridge, MA: Harvard University Press.

Sandborg, V. 1996. cited in "Baxter looking at ISO 14000 but favors own system." *Integrated Management System Update,* June: 23.

Sandholtz, W. 1999. Globalization and the evolution of rules. In A. Prakash and J. A. Hart (eds.), *Globalization and Governance,* London and New York: Routledge.

Sarkar, S. 1989. *Modern India 1885–1947.* New York: St. Martin's Press.

Sarokin, D. 1999. Email conversation. February 11.

Schall, L.D., Sundem, G. L., and Geijsbeek, W. R. 1978. Survey and analysis of capital budgeting methods. *Journal of Finance,* 33: 281–87.

Schelling, T. 1978. *Micromotives and Macrobehavior.* New York: Norton.

Schmidheiny, S. in collaboration with the Business Council for Sustainable Development. 1992. *Changing Course: A Global Business Perspective on Environment and Development.* Cambridge, MA: MIT Press.

Schmidheiny, S. and Zorraquin, F.J.L. (in collaboration with the Business Council for Sustainable Development). 1996. *Financing Change.* Cambridge, MA: MIT Press.

Schneider, M. and Teske, P. with Mintrom, M. 1995. *Public Entrepreneur: Agents for Change in American Government.* Princeton, NJ: Princeton University Press.

Scott, W. R. 1987. The adolescence of institutional theory. *Administrative Science Quarterly,* 32: 493–511.

Selznick, P. 1957. *Leadership in Administration*. Evanston, IL: Row, Peterson, and Company.

Senge, P. M. 1994. *The Fifth Discipline: The Area and Practice of the Learning Organization*. New York: Doubleday/Currency.

Shrivastava, P. 1995. The role of corporation in achieving ecological sustainability. *Academy of Management Review*, 20: 936–60.

Simon, H. A. 1957. *Models of Man*. New York: John Wiley.

Simmons, P. and Wynne, B. 1993. Responsible Care: Trust, credibility, and environmental management. In K. Fischer and J. Schot (eds.), *Environmental Strategies for Industry*. Washington, DC: Island Press, pp. 201–26.

Smith, A. 1970. *The Wealth of Nations: Books I–III*. London: Pelican Books.

Snidal, D. 1995. The politics of scope: Endogenous actors, heterogeneity, and institutions. In R. O. Keohane and E. Ostrom (eds.), *Local Commons and Global Interdependence*. London: Sage, pp. 47–70.

Stigler, G. 1971. The economic theory of regulation. *Bell Journal of Economics*, 2: 3–21.

Thompson, A. R. 1973. *Economics of Firm*. Engelwood Cliffs, NJ: Prentice Hall.

Tiebout, C. M. 1956. A pure theory of local expenditure. *Journal of Political Economy*, 64: 416–24.

Tietenberg, T. H. 1992. *Environmental and Natural Resource Economics*. New York: Harper Collins.

Tolbert, P. S. 1985. Resource dependence and institutional environment: Sources of administrative structures in institutions of higher education. *Administrative Science Quarterly*, 30: 1–13.

United Nations Conference on Trade and Development/UNCTAD. 1993. *Environmental Management in Transnational Corporations: Report on the Benchmark Corporate Environmental Survey*. New York and Geneva: United Nations.

1997, *World Investment Report*, 1997. New York and Geneva: United Nations.

US Environmental Protection Agency (USEPA). 1987. Underground storage tanks: Technical requirement. 52 *Federal Register* 12662, 12664, April 7.

1988a Underground storage tanks containing hazardous substances: Financial responsibility requirement, advance notice of proposed rulemaking. 53 *Federal Register* 3817, 3819, February 9.

1988b. Technical requirement: Final rule. 53 *Federal Register* 37082, 37088, September 23.

Vangeloff, J. 1992. Above ground small volume solvent storage. In *Above Ground storage tanks*, Proceedings of the Second International Symposium on Above Ground Tanks, January 14–16, 1992, Houston, Texas, pp.5/1–5/8.

Vogel, D. 1986. *National Styles of Regulation*. Ithaca, NY: Cornell University Press.

1995. *Trading Up*. Cambridge, MA: Harvard University Press.

Walters, I. 1973. The pollution content of American trade. *Western Economic Journal*, 11: 61–70.

Walley, N. and Whitehead, B. 1994. It's not easy being green. *Harvard Business Review*, May–June: 46–52.

Wartick, S.L. and Cochran, P.L. 1985. The evolution of the corporate social performance model. *Academy of Management Review* 4: 758–69.

Weber, M. 1947. *The Theory of Social and Economic Organization.* New York: Free Press.

Weick, K. E. 1995. *Sensemaking in Organizations.* London: Sage.

Weimer, D. L. (ed.) 1997. *The Political Economy of Property Rights.* Cambridge University Press.

Weimer, D. L. and Vining, A.R. 1992. *Policy Analysis: Concepts and Practice.* Englewood Cliffs, NJ: Prentice Hall.

Wendt, A. 1992. Anarchy is what the states make of it: The social construction of power politics. *International Organization,* 46: 391–425.

Wicks, M. E. 1980. *A Comparative Analysis of Foreign Investment Evaluations Practices of US Based Multinational Corporations.* New York: McKinsey and Company.

Wilkins, J. R. 1996 Lilly fulfills pledge to reduce emission, A note prepared for submission to the EPA. July 2, 1996. Indianapolis, IN: Eli Lilly and Company.

Williamson, O. E. 1964. *The Economics of Discretionary Behavior.* Englewood Cliffs, NJ: Prentice Hall.

1975 *Market and Hierarchies: Analysis and Antitrust Implications.* New York: Free Press.

1985. *Economic Institutions of Capitalism.* New York: Free Press.

Wiseman, J. 1957. The theory of public utility price – An empty box. *Oxford Economic Papers,* 9: 56–74.

Wood, D.J. 1991. Corporate social performance revisited. *Academy of Management Review,* 16: 691–718.

Wood, D.J. and Jones, R.E. 1995. Stakeholder mismatching: A theoretical problem in empirical research on corporate social performance. *The International Journal of Organizational Analysis,* 3: 229–67.

World Bank. 1992. *World Development Report.* Washington, DC: World Bank; Oxford: Oxford University Press.

World Commission on Environment and Development. 1987. *Our Common Future.* Oxford: Oxford University Press.

Yukul, G. A. 1981. *Leadership in Organization.* Englewood Cliffs, NJ: Prentice Hall.

Zinn, H. 1995. *A People's History of the United States: 1492 to Present.* New York: Harper Perennial.

Zucker, L. G. (ed.) 1988. *Institutional Patterns and Organization.* New York: Ballinger.

Name index

Abbey, G. M., 36, 38, 39, 42, 68, 81, 100, 112, 113, 155, 159
Abernathy, W., 160
Ackerman, R. W., 5
Agle, B. R., 28, 148
Alchian, A. A., 24
Allegiance, 11, 35
Allison, G. T., 2, 32, 144
Ambrose, S. E., 146
Appleby, J. O., 146
Arkansas Gazette, 79
Arora, S., 4
Arrow, K. J., 16, 18
Arthur D. Little, Inc., 38, 94, 109, 119, 123, 137, 145, 155
Atkinson, S., 1
Ayman, R., 30

Babcock, J., 110
Baker, D., 102
Barnard, C. I., 30, 31
Barrett, S., 4
Barton, S. L., 5
Bass, B. M., 30
Bates, R. H., 16
Baumol, W. J., 1, 25
Bayh, Governor, 48
Berle, A. A., 6, 19
Bernstein, M. H., 117, 147
Bessen, S. M., 123
Blackburn, W. R., 38–40, 68, 69, 81, 123
Booz, Allen, and Hamilton, 42
Boulding, K. E., 30, 31
Bowles, S., 29
Bowman, D., 64
Brannon, D., 48, 49
Browner, C., 106
Buchanan, J. M., 18, 22
Bunge, J., 2
Bushouse, B., 22
Buzelli, D., 83

Carmichael, D., 68, 111
Carroll, A. B., 5
Carson, R., 1
Cascio, J., 117
Casson, T. N., 4
Chandler, A., 28, 67, 104
Charter, M., 95
Chatterjee, P., 85
Chemers, M. M., 30
Chemical Manufacturers
    Association/CMA, 22, 71, 83–9, 114, 121, 150, 156
Child, J., 6
Choucri, N., 9
Clarkson, M. B. E., 5
Coase, R. H., 7, 27, 139, 157
Cochran, P. L., 5
Coddington, W., 92
Cohen, M. D., 156
Cohen-Rosenthal, E., 2
Colborn, T., 90, 101
Commons, J. R., 22
Cooper and Lybrand, 42
Cornes, R., 22
Crandle, 68
Crawford, S. E. S., 147
Cross, J. G., 20
Currie, R. M., 73, 75, 82
Cyert, R. M., 2, 8, 65

Dahl, R. A., 28
Damon, L. A., 4
Datta, K., 146
Davis, T. R. V., 30
Deming, W. E., 114
Demsetz, H., 19
DiMaggio, P., 2
Dodd, L., 32
Donaldson, G., 160
Donaldson, T., 5
Dow, G., 28

Subject index

adversary economy, 67, 104, 117
agent-structure, 5–9, 17, 19
Allegiance Corporation, 11
American Hospital Supply Corporation
  (AHSC), 37, 62, 81
Arthur D. Little, 38, 94, 109, 119, 123, 137,
  145, 155

Baxter
  Awards, 125–33
  Bentley Laboratories, 77, 109
  Caremark, 102
  Corporate Environmental Affairs, 40, 43,
    68
  Corporate Environmental Engineering,
    36, 40, 63
  Corporate Environmental Law, 38
  Customer Waste Council, 97
  Divisional Environmental Managers,
    42–4, 53, 145
  Environmental Review Board, 36, 38–40,
    42–3, 53, 63, 68, 100
  Environmental Financial Statement, 41
  Facility Environmental Managers, 42–3
  Manford facility, 78
  Newsletters and Reports, 37, 63, 69, 82,
    102, 109
  Overview, 10–11, 34–5
  V. Mueller, 78
behavioral theory, 2, 165
beyond-compliance policies, 1–6, 152–3
Bhopal, 71, 84, 106
Booz Allen and Hamilton, 42
bounded rationality, 8, 19–20, 24

capital-budgeting, 3, 6–7, 23–7, 33, 136–8,
  160–1
  IRR, 25
  NPV, 25
capture theories, 117, 147
case selection, 13–14

Chemical Manufacturers Association
  history, 83
  Management System Verification, 50,
    87
  policies, 71
  Responsible Care, 12, 22–3, 48, 83–91,
    121–2, 124–5, 150–1, 156
club goods, 20–3
Coalition of North Eastern Governors
  (CONEG), 40, 99
command and control policies, 1–2
common pool resources, 20–2
Cooper and Lybrand, 42
core competencies, 45
corporate social performance, 5

DEHP, 101
dependent variable, 12, 14
discursive struggles, 8, 138
Dow-Canada, 83
Dow-Elanco, 90

efficiency
  intended, 24
  substantive, 23–4
  procedural, 23–4, 33
environmental audits
  external audits, 12, 38–40, 104–5, 115–16
  overview, 103–4
  internal audits, 12, 87, 104–5
environmental insurance, 67
environmental laws and programs
  Clean Air Act Amendments (CAAA), 48,
    81, 105
  Comprehensive Environmental
    Response, Compensation, and
    Liability Act (CERCLA), 47
  Emergency Planning and Community
    Right to Know Act (EPCRA), 1
  Hazardous and Solid Waste
    Amendments (HSWA), 61

THE BEST OF JOHN ADAIR ON LEADERSHIP AND MANAGEMENT

EDITED BY NEIL THOMAS

THOROGOOD

Thorogood Publishing
10-12 Rivington Street
London EC2A 3DU
Telephone: 020 7749 4748
Fax: 020 7729 6110
Email: info@thorogoodpublishing.co.uk
Web: www.thorogoodpublishing.co.uk

A CIP catalogue record for this book is
available from the British Library.

ISBN 1 85418 608 0
    978-185418608-9

Designed by Driftdesign

Printed in Great Britain by
Ashford Colour Press.

Foreword

Thanks to the excellent editorship of Neil Thomas this book contains all the essentials of my thinking on leadership and management. It is with great pleasure, then, that I welcome you to these pages. You will find in them a wide range of practical suggestions, frameworks and examples that are designed with one simple aim in view: *to help you to become a more effective leader and manager.*

To that end Neil has divided my practical guidance on effective leadership, teambuilding, creativity and innovation, problem solving, motivation and communication – in a nutshell, 'the human side of enterprise' – into two parts: *Part 1 Self-management* and *Part 2 Managing others.* That makes sense to me. After all, if you can't manage yourself how can you manage anyone else?

At the core of my philosophy is the three-circles model. What the model does in a simple (but not simplistic) and visual form is to show you the generic role of leader, namely achieving the task, building the team and developing the individual. It tells you what you are there to do. Those broad and overlapping functions are then further expressed as eight more specific functions. You will notice that this foundational framework also serves to integrate into one the substance of both leadership and managership, as two eyes are joined in sight.

Remember, however, that it is only a foundation – a basis upon which to build your own house of leadership and management. The way to grow as a leader - so that you come to fulfil your own personal vision of what a leader should be, know and do – is to take time out from your busy life at regular intervals for reflection. If you don't believe you have time to spare for such self-development then read my section on time management without further delay!

Reflection needs something to work upon. You will, I trust, find the frameworks, charts, diagrams and forms in these pages useful tools to aid you in that essential work of thoughtful evaluation. It is when the sparks of understanding jump both ways between the principles in this book and your own experience that you will be learning to lead and manage more effectively. You will then be on your way. For learning, of course, is a journey. It is one which includes drawing lessons from the things that don't go so well for you, as well as your achievements and successes. That can be tough – I speak from experience.

Search my many books and not once will you find that I say that leadership is easy. So becoming an effective leader is not easy either. The more time and effort that you put into studying and applying the principles in this book the more you will find that being a leader and manager comes naturally to you.

Through accepting and developing your talent for leadership in this way you will increase what you have to give to others and to society. What the world needs now is what your organisation is also seeking – 'good leaders and leaders for good' at all levels.

Besides practical guidance I hope that you will also find in these pages some wells of inspiration to keep you going forward when the journey is uphill. Drink from those waters and then you will in your turn – by your words of encouragement and, above all, by your example – be able to inspire others.

I shall be with you in spirit.

John Adair

Contents

PART ONE:
SELF-MANAGEMENT

one

01

TIME MANAGEMENT

01

Introduction

Time management is about managing your time with a focus on achievement: of doing and completing those things which you want to do and which need doing.

Time management is goal-driven and results oriented. Success in time management is measured by the quality of both your work and your personal life.

Tempus Fugit

Whilst it is true to say that life only makes sense in retrospect, it can be shaped by your sense of time and purpose. In keeping with business planning, *time* planning – and your approach to the use of your time (and to the extent that you can influence it, how others spend their time) – should be to avoid the trap of failing to plan, which is, in effect, planning to fail. In other words, if time is money, spend it wisely.

Basic approach to time management

You need to be certain that you:

- can define your business role and know what constitutes a successful outcome
- spend time thinking and planning for yourself and others
- have a clear understanding of your business purpose
- know the balance you wish to achieve between your business and your private commitments (and can identify the time demands on both).

The Adair ten principles of time management

1 Develop a personal sense of time

2 Identify long-term goals

3 Make medium-term plans

4 Plan the day

5 Make the **best** use of your **best** time

6 Organise office work

7 Manage meetings

8 Delegate effectively

9 Make use of committed time

10 Manage your health

At all costs you should avoid falling into one of the following stereotypes:

- a poor delegator
- a bad organiser
- an excellent procrastinator
- a poor performer at meetings
- a purposeless executive.

Developing a personal sense of time

First audit how you spend your time, then analyse how you can improve your use of time.

Time audit

Keep a record (a daily time log) of where your time currently goes – break your day into fifteen minutes chunks for recording purposes. Do this for a week or so and review after each batch of three or four days.

Time Log

	1 Lunch ○
	1300
8	**2** Called to MD's office –
0800	1400 Spoke to Val re holiday ○
	Booking holiday by phone
	Working on next years budget
9 Opened post	↓
0900 Discussed club outing	**3** Made telephone calls ○
John	1500
Jan - Dictation	↓
↓	Last minute letter for
10 Prepared for meeting	**4** post. Read article in
1000	1600 press. Talked to George re
	stock plan
Meeting with client	**5** Call from client chasing
11	1700 order. Chasing above order ○
1100	Started meeting report
↓	↓
Dealt with messages	**6** Went Home
12	1800
1200 ↓	○
Went to Lunch	

Subject	Time Spent
Planned meetings inc preparation	$\frac{1}{2}$ hrs
Unplanned / Informal meetings	$1\frac{1}{2}$ hrs
Letters / messages / trouble shooting	$2\frac{1}{2}$ hrs
Telephone	1 hr
Personal	$\frac{1}{2}$ hr
Writing	$\frac{1}{2}$ hr
Reading	$\frac{1}{4}$ hr
Planning	$\frac{1}{2}$ hr
Lunch	$1\frac{1}{4}$ hr

Keyline 10915

Reproduced from Keytime Time Management
with the kind permission of Keytime Management Development

Peter Drucker's view is that only when we can manage time can we manage anything. In managing time we first need to know how we use it now and then change what and when we do things. Your time audit will probably confirm the findings of an IBM research which showed that the four activities that take up over 50% of the average executive's time are:

1   Meetings
2   Reading and writing business materials
3   Telephoning
4   Travelling.

Your time audit can identify these and others by using symbols ascribed to activities, for example:

M	Meetings (in committee form)
Mi	Meetings (in one-to-one interview form)
F	Finance and figure work
T	Telephone
E	Email
Wr	Writing (reports)
Wd	Writing letters or dictation
T	Travelling
R&D	Research and development (including reading, internet research, training and thinking)
AOB	Any other business activity (should be specified)

Your time log can then be summarised in the following format:

Activity	Time Spent (in hours)	% of time	Comment (how to save time from now on)

Research indicates that we make assumptions about where our time goes and over-estimate time spent on telephone calls, correspondence, report writing and planning, but under-estimate time spent on email and in meetings and one-to-one discussions. Keeping a record will confirm how you really spend your time and enable you to change how you spend it.

Analyse and improve your use of time

What elements can you readily identify which you can immediately change? Experience shows that improvements lie in changing the way you handle:

- interruptions (in person or by the telephone or email)
- meetings
- travel
- incoming/outgoing mail and email.

You can improve your use of time if you ensure that:

1 your time is spent according to a clear idea of your priorities and main responsibilities

2 you isolate the unimportant and ruthlessly prune out unnecessary or unproductive activities

3 you combine any 'free' time (ie free from meetings or other people's demands) to create meaningful and usable time of your own

4 tasks are simplified where others would not be adversely affected

5 you are not doing tasks which could be performed by others.

The balance of this chapter looks at how to ensure you improve your time management. The approach taken is to work from the long-term back to the immediate future, analysing your goals and giving time management tips on how to achieve them. (The approach works for both business and personal time management.)

Identifying long-term goals

First of all, it is necessary to define your organisation's purpose and the purpose or your job, ie to what end is your time being expended.

Then, long-term goals can be set in terms of the results that the organisation wishes to achieve (and your role as part of those goals being achieved).

Defining the purpose of your organisation requires an answer to the basic question: why does this organisation exist? You should be able to write this business purpose down:

_____

_____

_____

_____

_____

_____

Defining the purpose of your own job requires an answer to the question: why does my job exist? Again, you should be able to write this purpose down:

_____

_____

_____

_____

_____

_____

Identifying long-term goals, the strategy of your business and your part in it, will result from pondering these questions:

Where	are we now? do we want to be in 3 or 5 year's time?
What	strengths and weaknesses do we have?
How	can we improve? can we get to where we want to be?

These same questions can be applied to your personal life.

The answers to these questions will help you identify long-term goals. The longer the timeframe the more fuzzy the goals become, so you should then reduce your field of vision to focus on tangible, attainable, definable and measurable goals, but not lose sight of the far ground.

You should reach a point where you can be clear about long-term aims/directions and medium, or short-term goals/objectives which will be met and which will be part of a plan to continue on the road of achieving your longer-term aspirations.

Making medium-term plans

Your key areas of responsibility (and how your performance will be measured) should be listed and for each you must set objectives with time budgeted for each.

	Key area of responsibility	Objectives for each area of responsibility	Time budgeted to achieve each objective	Review of objective achievements
1	a b c	a b c	a b c	a b c
2	a b c	a b c	a b c	a b c

3

a

b

c

a	a	a
b	b	b
c	c	c

4

a

b

c

a	a	a
b	b	b
c	c	c

5

a

b

c

a	a	a
b	b	b
c	c	c

The review of objective achievement (the measure of your performance) should be at the intervals you have budgeted for each (eg 3, 6 or 12 months).

Smarter objectives

As a test of your objective-setting skills, remember they need to be **Smarter**, to the power of two, ie **Smarter**[2] :

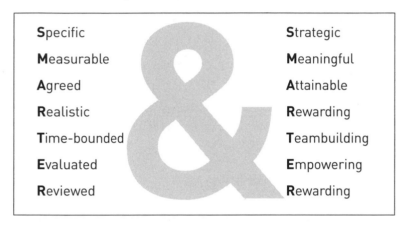

Specific	**S**trategic
Measurable	**M**eaningful
Agreed	**A**ttainable
Realistic	**R**ewarding
Time-bounded	**T**eambuilding
Evaluated	**E**mpowering
Reviewed	**R**ewarding

Part of all this is to set out clear ways in which time management can be improved in the short to medium-term. A 'Time Norm' form can help here.

Task/activity/procedure	Time taken now	Target time

When measuring and assessing improvements you cannot lose sight of the cost and quality dimension. Time improvements should not compromise standards set for those elements. Real improvement comes from keeping all three at whatever is decided are the acceptable levels.

Being successful in making medium-term plans requires you to:

- know the context (the longer-term) in which you operate and how the medium and longer-term goals are linked
- be able to plan and implement activity
- set clear objectives and review progress toward them on a regular basis
- be flexible and adaptable to change in order to stay on course to meet objectives (unless you have changed those, too!)

Planning the day

The golden rule is to plan an outline for each day a week ahead, but plan for the day in detail the day/evening before it, or at the beginning of it.

In setting your programme for the day you need to establish **priorities** related to **urgency** and **importance**. If you then spend time according to how you have set your priorities, you will have addressed the important jobs – that is the art of time management.

The Adair urgency/importance matrix

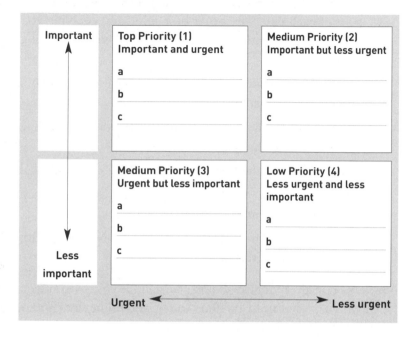

In the matrix you can identify tasks to:

1 do now

2 plan for (to use quality time)

3 do quickly (not requiring quality time)

4 do later or perhaps delegate.

This approach has also been called the Four-D system:

Drop it, Delay it, Delegate it or Do it.

You should shape your plan for the day by listing the various components, prioritising them and planning the time accordingly.

Your plan of action for the day should follow these rules:

- Make your plan at the end of the previous day or at the start of each day (whichever best suits you) enabling you to assess any unfinished work, together with upcoming priorities

- List the main elements (in relation to yesterday's, today's and the week's plans)

- Prioritise those main elements and identify tasks according to the matrix 1, 2, 3 and 4 above

- Group items together (eg telephone calls, correspondence)

- Decide when you will do the top priority tasks and block time out to do them

- Decide on remaining tasks (and share your plan with assistants/staff as relevant).

At a very basic level your list should also include your own system for identifying what **must** be done today, **should** be done today and what **might** be done today.

During the course of the day, regularly ask yourself whether you have changed priorities deliberately or whether you need to get back on course and tackle what remain as being prioritised activities.

Think of any daily list as a kind of shopping list – how are you going to feel going home without an item being ticked off as done?

However, do not be fixed and inflexible, provided you have managed your time and time has not mismanaged you.

Tips on daily planning

Tips on day-to-day planning and your programme centre on whether you are achieving at least your main priority activities.

If you are not, then consider:

- Are you unrealistic in trying to do too much?
- Were you unprepared (ie not ready) at the time you set to do the task?
- Was the task ill-defined?
- Did you find decision-making difficult (generally or for the task in question)?
- Were you lacking certain information?
- Is the neglect caused by too much time pressure?
- Did you 'give up' because of difficulty or boredom?

You should test whether your time budget made sense to start with, but if it did, then experience shows that the most common problem is interruptions. The most successful solution to interruptions is to learn to say 'No', remembering at the same time to be ruthless with time, but gracious with people.

Making the best use of your best time

You have certain times of the day when you perform consistently better than at other times. The Pareto Principle (which states that **significant** items of a given group form a relatively small part of the **total**) applies and 20% of your time produces 80% of your best quality output.

In making the best use of your best time, ie when you think straighter and are more alert, you need to know which part of your typical day it is that you can best perform certain types of task. Decide and plan activities accordingly.

You should be able to answer these questions. Are you:

- a morning, afternoon or evening person?
- aware that short-term memory is better in the morning?
- using the morning for words and figure work, when most people perform such work better than at other times?
- aware that we reach peak alertness at noon, or that we are manually more dextrous in the afternoon?

Planning breaks into your day (for coffee, lunch, walking around the office/factory to meet people) will help you to work in concentrated bursts. Creativity as well as in-depth work can be improved by managing your best times to do, or think about, the appropriate task. Be selfish about when and how you spend time on planned activities. Do not be afraid to work at unusual times if you find that it suits you, does not inconvenience others and is more productive.

Organising office work

The two key consumers of time are interruptions and paperwork. To learn how to be the victor rather than victim of these two thieves of time, you should follow these guidelines:

Dealing with interruptions

- Meet people in their office whenever you can (you control your leaving time)
- Stand rather than sit for casual visitors (this controls length of stay)
- Keep a focus on time (mention the time you have available, refer to your next meeting and have a visible and watched clock)
- Stick to the point and avoid butterflying from main topic to unrelated ones
- Be firm in a pleasant way.

Dealing with paperwork/email

- Do you see only what you should?
- Do you remove extraneous items?
- Do you handle items only once? (This one idea is known to save up to one hour per day or 220 hours a year!)
- Do you prioritise items (into action, information, reading, or for the wastebin/trash)?
- Do you limit the amount of paperwork and emails that you generate for others?
- Can you pick out salient points quickly and know when to skip read or read in-depth?

Other elements to improve your time management in the office

1 Arrange your office or office space for ease of work, comfort and efficiency. Few people give this any thought at all.

2 Clear desk policy – concentration is helped by doing one thing at a time so your desk should only have on it, the specific job that you are tackling at the time.

3 Write effectively, keeping it short and simple by thinking of the main point first and ordering your thoughts for logical expression.

4 Telephone – keep a log to see how time efficient you are now! Then get used to **planning** for each call you make (the salient points you want to make); **grouping** incoming and outgoing calls (usually for the end of the day when people are less verbose); and use a timer (eg an egg timer – to keep all calls to a maximum of four minutes). Do not be afraid to put a block on incoming calls to reduce interruptions.

5 Email – don't send them unless they are absolutely necessary; indicate clearly where action is required or where they are for information only; handle items at a time to suit you; keep in-box for current items only; handle spam so that it does not swamp you; and do not use if a telephone call would be better.

6 If you have an assistant, use him or her to deal with or to redirect (helpfully) any mail or callers (whether in person or on the telephone), where he/she or someone else could better deal with them. Strive for excellence not perfection, through your assistant.

Managing meetings

A Time Lord's approach to meetings is to confront three main issues:

- Is the meeting strictly necessary at all?
- How much time (particularly mine) is it worth?
- Will it run to time?

You must always have a clear idea of how much a meeting costs (in people's time, including your own) and whether it is worth it in results. For example, typical costs might be:

Salary p.a.	Per hour cost	Per day cost
£100,000	£60	£420
£50,000	£30	£210
£25,000	£15	£105
£12,500	£7.50	£52.50

(These figures are only based on salary levels and exclude over-heads, to say nothing of the opportunity costs of attending meetings rather than spending time on other productive tasks.)

What then are the hallmarks of the successful manager of meetings?

- Meetings are planned ahead (who should attend and with the agenda and any useful papers being circulated in advance)
- Times for each item and of the meeting itself are set in advance (and adhered to)
- Minutes are concise and action-oriented (with responsibilities allocated)
- There is clarity of outcomes(s) (shared by all)

- Meetings are reviewed constantly for effectiveness
- The focus is on the positive
- You are a successful umpire and referee.

Before holding any meeting, ask yourself these five questions:

1 *Why* are we meeting?
2 *What* would be the result of not having the meeting, or *what* should result from having it?
3 *Who* should attend?
4 *How* long should it be and *how* should it be structured?
5 *When* is the best time to hold it?

You cannot ban all meetings, so you must manage them to get the best results. To do this you need to identify the type of meeting:

The Adair five types of meeting

1 **Briefing meetings**
to impart and share information, to clarify points and incorporate ideas from others

2 **Advisory meetings**
to gather views and advice and to outline or share any ideas

3 **'Council' meetings**
to make and share responsibility for decisions, resolving differences on the way

4 **Committee meetings**
to 'vote' decisions and reach compromises/ accommodations of different views on matters of common concern

5 **Negotiating meetings**
to reach decisions by bargaining with other party(ies) who are acting in their own best interest

You should decide what each type of meeting you are to be involved with actually is and plan to run each type as time efficiently as possible depending on their purpose.

Being aware of the cost of meetings will focus the mind and planning will focus your actions. Minutes to record actions agreed and responsibilities should be in a form to give ease of follow-up and subsequent checking.

Delegating effectively

Delegation is entrusting a job to another who is also given the authority to do it. It is vital to be clear that delegation is not abdication – to give up your power to another would be an evasion of responsibility.

The benefits of delegation are that it gives you more time to carry out your key functions and to develop your subordinates. You are freed to spend more time on management and leadership and you are able to concentrate on such areas as:

- strategic and development issues
- knotty problems
- staff appointments/development/training
- key marketing/quality issues
- emerging from your office to hear what staff/customers can tell you about your business
- communicating with more of your staff.

Deciding what to delegate

You should select the type of work for delegation and consider to whom it can be delegated.

Type of work	Delegation to enact
1 Technical/specialist work	
a _____	a _____
b _____	b _____
c _____	c _____
2 Administrative/minor decisions	
a _____	a _____
b _____	b _____
c _____	c _____
3 Where others are more qualified	
a _____	a _____
b _____	b _____
c _____	c _____
4 Where staff development would result	
a _____	a _____
b _____	b _____
c _____	c _____

The seven main reasons why CEO's do not delegate were revealed by research in five European countries to be:

1 It is risky

2 We enjoy doing things

3 We dare not sit and think

4 It is a slow process

5 We like to be 'on top of everything'

6 Will our subordinates outstrip us?

7 'Nobody can do it as well as I can'.

So, what qualities must you have to be a good delegator?

There are five main tips:

1 Choose the right staff

2 Train them

3 Take care in briefing them, and ensuring their understanding of the why and 'how to' of tasks delegated to them (and in imparting to them an understanding of business aims and policies)

4 Try not to interfere – stand back and support

5 Control in a sensible and sensitive manner by checking progress at agreed intervals.

A checklist to test if you are an effective delegator

	YES	NO
Do you take work home evenings/weekends and/or work more than nine hours a day?	☐	☐
Can you identify areas of work that you could/should delegate, but have not already done so?	☐	☐
Do you define clearly the delegated tasks and satisfy yourself that the individual to whom they are delegated understands what is expected as an outcome?	☐	☐
Can you trust people, or do you find it difficult to do so?	☐	☐
Do you delegate authority and tasks?	☐	☐
Do you think that the delegated task will not be done as well by anyone else?	☐	☐
Do you involve those to whom tasks have been delegated in the whole planning and problem-solving process?	☐	☐

Making use of committed time

You can increase your achievement level by using committed time (time that is 'booked' for example – travelling, meal-times) by: in the case of travelling time, ensuring that you use it to carry out, for example, reading, writing, thinking, meetings, making phone calls, listening to 'improvement' tapes; in the case of meal times, ensuring that you use them where relevant to hold business conversations or meetings.

In other words, you should establish productive activities to schedule alongside time which has to be committed to other activities, eg:

- Daily routines: use dressing, washing/shaving/meals etc by stimulating your mind, eg with an improving tape
- Waiting time: do not waste it
- Travel time: use it productively
- Television: Do not let it consume too much of every evening.

Managing your health

Time management has to be as much about ensuring that you maximise the amount of time you have available to use as well as using the time wisely. That means taking steps to ensure you do not suffer time deprivation through illness of mind or body.

It is vital to look after your energy levels – regularly to top up your batteries – to discharge efficiently in a long-life fashion.

How topped-up are your batteries?

This is a five point test:

Sleep are you getting enough
(*Guide* – eight hours or slightly less with age)

Eating habits – are you being sensible?
(*Guide* – moderation in all things)

Holidays – do you take and enjoy your full entitlement?

Exercise – honestly what are you doing each week?
(*Guide* – three sessions of 20 minutes per week – even brisk walks – is a reasonable aim)

Thinking time – do you think about what you are doing in life and in work, even for a few minutes each day?

Stress

Time and stress are cyclists on the same tandem. Bad management of one pushes the other out of control.

You should **always** watch out for signs of stress in yourself and others... and take corrective action.

Warning bells should sound if any of the following are present in behavioural patterns:

1 Irritability
2 Ever-present anxiety/worry
3 Constant tiredness
4 Increased consumption:
   – alcohol
   – cigarettes
   – drugs
5 Over-eating or the loss of appetite
6 Memory lapses
7 Loss of a sense of humour
8 Feelings of nausea/fainting spells
9 Poor sleep patterns
10 Nail-biting, nervous mannerisms/'tics'
11 Feelings of tension and headaches
12 Indigestion
13 Loss of concentration
14 Unable to relax
15 Feeling unable to cope
16 Indecision
17 Erratic driving
18 Dependence on sleeping pills
19 Sweating for no apparent reason
20 Frequently crying or wanting to cry.

Research (in ten countries on a thousand managers) reveals that improving time management can help eliminate the 12 most common roots of stress in managers which are:

1 Time pressures and deadlines
2 Work overload
3 Inadequately trained subordinates
4 Long working hours
5 Attending meetings
6 Demands of work on private and social life
7 Keeping up with new technology
8 Holding beliefs conflicting with those of the organisation
9 Taking work home

| **10** *Lack of power and influence* | **12** *Doing a job below one's level of competence.* |
| **11** *The amount of travel required by work* | |

If you find yourself suffering from stress then you must:

1 do something about it: look at the stress factors and assess what can be done to change your life at work/home.

2 express yourself: talk to people about how you are feeling and the concerns you have (even directly to a person who might be causing part of your stress).

3 evaluate priorities: check the balance of your life, take stock of activities and priorities and change them if necessary.

4 accept what you cannot control: have the courage to change the things that can be changed, the serenity to accept the things that cannot be changed and the wisdom to know the difference.

5 use your negative experience to positively change your behaviour.

6 use time management skills to take charge of your time and how it is spent, particularly making time to deal with stress causing problems. Get them out of the way.

7 count your blessings – list those things that you are pleased with, about yourself or achievements. Do not over concentrate on the past (guilt) or the future (anxiety).

8 ask yourself – what is the worst that can happen and can I cope with that? Use this to reduce anxiety about an issue.

Summary and six-month follow-up test

Keep the Adair ten principles of time management in the forefront of your mind and in your planning and prioritising:

The Adair ten principles of time management

1 Develop a personal sense of time

2 Identify long-term goals

3 Make medium-term plans

4 Plan the day

5 Make the **best** use of your **best** time

6 Organise office work

7 Manage meetings

8 Delegate effectively

9 Make use of committed time

10 Manage your health

Six-month follow-up test

You should periodically test your time management skills by asking yourself these questions:

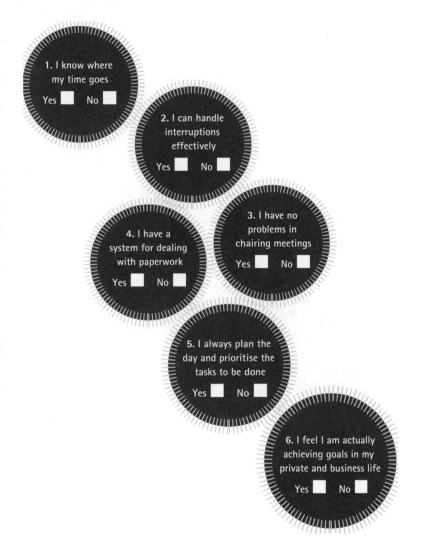

1. I know where my time goes

Yes ☐ No ☐

2. I can handle interruptions effectively

Yes ☐ No ☐

3. I have no problems in chairing meetings

Yes ☐ No ☐

4. I have a system for dealing with paperwork

Yes ☐ No ☐

5. I always plan the day and prioritise the tasks to be done

Yes ☐ No ☐

6. I feel I am actually achieving goals in my private and business life

Yes ☐ No ☐

02

SETTING AND ACHIEVING GOALS AND OBJECTIVES

02

Introduction

In the chapter on Time management, the importance of identifying goals and making plans was outlined in relation to managing your time, in order to achieve the goals/objectives that you set and where the progress toward them could then be measured.

What do the words 'goal' and 'objective' mean? Defining them can assist in setting your goals/objectives.

A **goal** is the end towards which effort or ambition is directed. As a word it has its origins in 'the point marking the end of a race' or 'posts between which a ball is to be driven'. Goal-setting is a successful way to approach life-planning.

An **objective** (a word with military associations) is an end towards which effort is directed.

So goals/objectives define the end or purpose which is being aimed for.

Goals and objectives can be used in your personal and business/professional life and successful outcomes result from taking a strategic approach to your life and your work.

In a strategic approach, your aspirations need to be fixed in the form of goals/objectives. Strategy is then the way of devising plans and using stratagems (the devices/means) towards achieving the goals/objectives. (Objective and strategy are both words with military connotations and origins.)

Personal goals and objectives

The starting point is to self-assess and take stock of yourself in the overall context of the direction you would like to be heading.

Personal Profile

1 What are my strengths/what special skills do I have?

_____

_____

_____

2 What are my values (ie what is important and worthwhile to me)?

_____

_____

_____

3 What would be my preferred ways of earning a living?

_____

_____

_____

4 What activities/situations do I want to avoid?

_____

_____

_____

5 What achievements would I like to list as having been successfully met in my life?

_____

_____

_____

6 What would I like the highlights of my obituary to be? (This is a really good way of focusing the mind, as is asking yourself how you would like your epitaph to read.)

_____

_____

_____

Answering the above questions will enable you to set out more clearly the goals/objectives you want to achieve in your life and, linked with your time management skills, to plan the important steps to take to achieve them. You will then be able to 'add years to your life and more life to your years'.

Again the approach is to work from the long-term back to the short-term as in:

1 What are my lifetime goals/objectives?

_____

_____

_____

2 What are my five-year goals/objectives?

_____

_____

_____

3 What goals/objectives will I set for achievement within one year?

_____

_____

_____

Taking this approach will ensure that you concentrate on those goals/objectives which are important to you. The strategic element of your approach will then ensure that you analyse the obstacles that have to be overcome and plan your priorities and the ways/ means to achieve your one year, five year and lifetime goals/objectives.

Answer these two 'questions' for each set of goals/objectives.

1 Identify obstacles and opportunities:

_____

_____

_____

2 What are the ways/means to achieve goals/objectives, despite/because of those obstacles/opportunities?

_____

_____

_____

This will give you your plan(s) of how to achieve your personal goals/objectives, which you must then implement!

Professional/business goals and objectives

You will find it helpful at this point to re-read the sections on **Identifying long-term goals** and **Making medium-term plans** in the section on Time management.

In setting and achieving professional/business goals and objectives you must:

- Define the purpose (of the business and of your role)
- Define your strategic aims

As has been seen, your strategic aims result from asking the questions:

Where   are we now?
             do we want to be in 3 or 5 year's time?

What   strengths and weaknesses do we have?

How   can we improve?
             can we get to where we want to be?

As for setting personal goals/objectives, your professional business goals/objectives (for your organisation and, perhaps separately, for yourself at work), should be delineated:

1 What are the key long-term goals/objectives?

_____

_____

_____

2 What are the five year goals/objectives?

_____

_____

_____

3 What are the one year goals/objectives?

_____

_____

_____

Then, plans/strategies have to be addressed as in:

1 Identifying obstacles and opportunities:

_____

_____

_____

2 Analysing ways/means to achieve goals/objectives despite/because of those obstacles/opportunities:

_____

_____

_____

Planning answers the question: '*How* are we going to achieve' a particular task, meet a goal or reach an objective? *How* leads to *who*, *what* and *when*? You can then set out your:

3 Strategy for achieving:

a) short-term goals/objectives – the one year:

_____

_____

_____

b) medium-term goals/objectives – the five years:

_____

_____

_____

c) the long, long-term goals/objectives:

_____

_____

_____

In the context of goals/objectives and strategy, there are a variety of different words/expressions used, as we have already demonstrated with the use of the words, **goals** and **objectives**. The ten key words in this area and what they suggest are:

1 **Goal:** the selected result to be achieved only with effort

2 **Objective:** tangible, attainable end toward which effort is expended

3 **Purpose:** the desired result or movement toward it

4 **Aim:** the target aimed at

5 **End:** the intended effects of actions (as opposed to means)

6 **Object:** the same as end, but also conveys wish and motive

7 **Mission:** the purpose for which the individual or organisation was 'sent'

8 **Plan:** set of ideas to achieve desired result

9 **Vision:** a concept denoting clarity of foresight/sharpness of understanding of a scheme

10 **Intention:** what is in your mind to do.

In order to reach the goals/objectives set, you will need to identify them clearly. The hallmarks of good goals/objectives (see also the **SMARTER**[2] test in the earlier section on Making medium-term plans) are as follows:

Goals/objectives must be:

Clear, specific, measurable, attainable, written, time-bounded, realistic, challenging, agreed, consistent, worthwhile and participative.

Attaining goals/objectives brings into play strategy and planning, for which you need imagination, a sense of reality, power of analysis and what has been described as helicopter vision (the ability to see matters in detail, but from a higher perspective).

An operational plan/strategy should contain:

- a **SMARTER**[2] set of goals/objectives
- plans/stratagems for achieving them.

The plan should detail all the steps required to complete those tasks which are needed to be done in order to meet the objectives set.

Time spent on planning is crucial in the thinking, the brainstorming and the sharing of ideas, the definition of purpose, the identification of obstacles and the ways and means of overcoming them and of meeting objectives.

It is necessary, of course, to get the right balance between **planning** and **implementation**. Planning saves time at the strategic and operational level and the key principle is 'every moment spent planning saves three or four in execution'.

Summary and six-month follow-up test

This chapter has sought to emphasise that in your business/ professional and your personal life you must:

- set goals/objectives (short, medium and long-term)
- create plan(s) to achieve them
- implement the plan(s) and achieve them.

Six-month follow-up test

1. Have you defined 'purpose' in your personal and business life?

Yes ☐ No ☐

2. Have you identified long-term goals?

Yes ☐ No ☐

3. Have you set goals and objectives?

Yes ☐ No ☐

4. Have you made short, medium and long-term plans?

Yes ☐ No ☐

5. Have you budgeted time to achieve goals/objectives

Yes ☐ No ☐

03

DECISION-MAKING AND PROBLEM-SOLVING

03

Introduction

Decision-making, along with leadership and communication is one of the top three attributes a successful manager needs. It is a direct result of 'thinking' and you need to be able to 'think until it hurts'.

Decision-making is directed to reaching a goal/objective. It is about the how, what, why, when (and where) of a course of action and of how to overcome obstacles and to solve problems. Decision-making is what turns thought into action: it implies change and requires a decision to be made against a background of uncertainty and risk.

(In this chapter the use of the term 'decision-making' embraces 'problem-solving'.)

Decision-making skills

You need to be able to choose the action or course of action that is the best for you/your organisation to meet its objective(s). An effective decision is one that produces the goods, ie gives the desired end result.

It is important to be able to project ahead, to take the expected and unexpected into account, to have contingency plans in case events intrude in such a way as will turn a good decision into a bad one.

There are usually several different decisions that can be taken and pressure to decide. Decide you must, even if trial and error are then used to assess the decision, amend it or overturn it.

Fear of failure must not serve to make you risk-averse, rather it should push you harder to 'think until it hurts'.

The effective decision has these six elements:

1 **Defining the objective**

2 **Gathering sufficient information**

3 **Identifying the feasible options**

4 **Evaluating those options**

5 **Making the decision (choosing an option)**

6 **Testing its implementation: by feel, by measurement and by assessment.**

You should also listen to your 'feel-right?' test – do warning lights flash or alarm bells sound? If so, re-work decision elements 1-6. (Experience of your own or that of others helps to develop your 'feel' for decisions).

A decision is only effective if it is implemented (and that means getting the desired results through people). For that, other people need to be included in the decision-making process. You need to develop your skills in appreciating when it is most appropriate to include others in the decision-making process.

An effective decision-maker is always an effective thinker. The three essential skills are those of:

1 analysing

2 synthesising

3 valuing.

An effective decision-maker knows that quick decisions are not necessarily the best ones and decisiveness only results from think-ing things through. Key decisions (and recognising when you are being asked to make or be involved in the making of key decisions) demand that great care must go into **analysing** (the component elements), **synthesising** (putting ideas together) and **valuing** (assessing relative worth).

The crucial elements in decision-making are:

- establish the facts
- consider the options
- decide the course of action.

The truly effective thinker has these attributes:

1 **Skills of analysis, synthesis and valuing**
2 **Open to intuition**
3 **Imagination (to find new ways to overcome problems)**
4 **Creativity (coupled with careful preliminary work)**
5 **Open to new ideas**
6 **Humility – when to recognise that others may have better powers or knowledge and to combine with their thinking.**

To improve your performance you need to ask yourself how good your skills are (and have been) at:

- decision-making
- problem-solving
- creative thinking.

Do you make false assumptions and jump to conclusions or are you prone to faulty reasoning or to not listening to others?

Two maxims are useful in decision-making:

- 'It can be done'
- 'Always try to turn a disadvantage into an advantage.'

Always operate in the context of facing reality and of seeking and speaking the truth.

The manager as decision-maker

Management can be said to be 'deciding what to do and getting it done'.

Success in business stems from good quality management decisions first of all and then the effectiveness in implementation which depends on the skills of leadership, influencing, communication and motivation.

One survey (of 200 leaders of industry and commerce) ranked 'the ability to take decisions' as the most important attribute of top management.

The logical or rational manager will invariably follow this decision-making model:

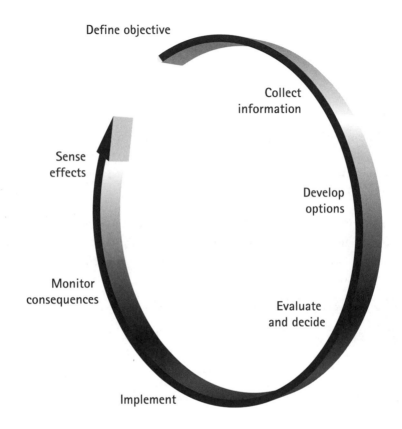

Define objective

Collect information

Sense effects

Develop options

Monitor consequences

Evaluate and decide

Implement

However, US research into decisions by public sector officials suggests that decision-makers rarely settle for the 'best' or optimum solution, being affected by emotion, power politics, the influences of other people and by their own values. Often a decision is a compromise between different courses of action, being one that:

- agrees to some extent with one's own personal interests, needs or values

- meets the value standards of superiors
- is acceptable to those affected (by the decision and for carrying it out)
- looks reasonable
- has an escape element of self-justification if it all goes wrong.

Clearly such approaches to decision-making must be removed from your approach!

Finally, managers need to be prepared to make time to think about decisions – to devote quality time to this crucial area of activity; to avoid superficiality (resulting from performing a great variety of tasks with little time to spare) because thought must be as important as action; and to involve other people (colleagues, subordinates and superiors) in making sensible management decisions.

Key elements of effective thinking and decision-making

Analysis

An essential ability in analysing is to be able to break the whole up into its component parts, particularly complex matters into its simple elements.

The hallmarks of the analytical mind are that it:

1 establishes the relationship between the parts and the whole
2 finds the root cause(s) of the problem
3 identifies the issue(s) at stake, the 'either/or' upon which a decision rests.

Analytical ability is improved by:

- working from first principles
- establishing the facts and separating them from opinions, assumptions or suppositions
- asking yourself questions (as in 'When did the problem first arise?', as well as our six friends Who, What, Why, When, Where and How?)
- constantly checking the premise and/or logical steps which can undermine good reasoning
- thinking backwards from the desired outcome
- organising the facts
- seeing the problem as a solution in disguise.

Analysis is not, however, an end in itself and trying to over-analyse can lead to inactivity or 'paralysis by analysis' as it has been called.

Synthesis

Decision-making requires an individual to 'take a view' and that depends on the ability to combine parts or elements to form a whole: synthesis. **Holistic** is a useful word to use in this regard as it also conveys the approach, especially in business, which recognises that 'wholes' are produced by grouping various units together where it is difficult to analyse them into their parts without losing this wholeness. Hence an holistic view needs to be taken in business decision-making.

One difficulty is that analysis can be the enemy of synthesis (holism) and vice versa. There is a need in business to be able to see the wood for the trees (holism) rather than only the trees (analysis).

In this sense, and in business too, the whole is greater than the sum of its parts. Business thinking is a good example of the Gestalt approach whereby we arrive at an understanding of:

- the overall significance rather than a mechanistic explanation
- the relationships between events not just the events themselves which do not occur in isolation, only in a setting which gives each significance.

Managers need to take this whole view – not to see things as a marketing problem, or a production issue, or a stock control difficulty, or a people problem, or a management failure. Look at the whole to see what that can yield by way of a solution.

Integration of facts, ideas and opinions is like the ability to synthesise and strengthens the manager's decision-making. Particularly in assessing financial performance, a manager needs to view the figures as a whole as well as in detail.

Other useful approaches

Imagination

This is an important attribute to have in business: the skill to visualise the whole in one's imagination. It is part and parcel of being creative in the approach to decision-making. Being imaginative is a crucial ability to develop in oneself and others: it helps to surprise the competition, to exploit the unexpected, to invent new products or services, or to solve problems creatively. Indicators of a healthy level of imagination are the abilities to:

- recall events easily and visually
- foresee what may happen before an event happens

- be inventive or creative artistically, mechanically or verbally
- fantasise about future events.

These elements of recall, visualising, creating, foreseeing and fanta-sising contribute to effective thinking in business as much as in the arts or scientific fields.

Conceptual thinking

Although a **concept** may appear to be an abstraction arrived at by analysis, it has a different feel because:

1 **it is a whole (and as such more than the sum of its parts); and**
2 **it is a developing entity in its own right.**

A concept is 'something conceived in the mind' and conceptual thinking in business addresses such issues as:

- What business are we in?
- What are its strengths/weaknesses?
- What are its purposes/aims?

Conceptual thinking should be kept separate from decision-making, even though decisions are made on the basis of the concepts that we have.

Concepts can be used in 'profiling' business development, but they then have to be made more specific in the form of proposals or plans, before being implemented. Concepts can be a way of taking your mind away from the particular and include the ideas of what ought to be as opposed to what is. Good quality concepts will under-pin good quality business decisions. Therefore you should generate clear, well-defined concepts and develop them.

Intuition

Being intuitive, successfully so, is undoubtedly a help in making effective decisions. It is not always possible to analyse problems into solutions and intuition is the useful power to know what has happened or what to do. Interestingly the powers of intuition are diminished by stress and general fatigue, and so your ability to be insightful in decision-making can be adversely affected by these factors.

'Intuition', 'instinct', 'first impressions', 'feel', 'hunch' and 'flair' are important dimensions to explore when faced with not only decision-making, but also considering business activities and the systems to run them.

It is too easy to be dismissive of intuition, of being able to 'sense' what needs to be done or to 'smell' trouble/opportunities. Rather it is an invaluable key to making and taking effective decisions.

Originality and innovation

Creative and innovative thinking can help in making decisions that develop a business so they are elements to encourage in yourself and others. Be prepared to work at problems/issues to encourage creativity or insight coming into play and be prepared to encourage new ideas (by rewarding those who put them forward), to try out and innovate new products/services as well as new ways of doing things.

The concept of value in decision-making

With analysis and synthesis, valuing is the third essential in effective thinking and decision-making. The ability to make decisions has two main aspects:

1 **To establish the truth or true facts; and**

2 **To know what to do.**

Time must be taken on the first, otherwise integrity, or the value of truth, is lost in the process. Thinking first and then deciding what to do is the correct order in decision-making. Getting at the truth should make knowing what to do easier.

In many respects, it is better to behave as if truth is an object, that it must be discovered. The truth and valuing that one discovers, should be seen as 'objective' with one's own views and conditioning recognised and relied on or discounted as needs be.

When you rely on others, as managers so often do, you may have to sift information from their 'valuations' (information plus judgement). This is another form of valuing – of knowing who and when to trust to give you truth, or truth backed by acceptable value judgements. Questioning is a valid part of establishing the credentials of the adviser and the credibility of the advice. Can you trust the person to tell the truth backed by sufficient expertise or insight? You will learn by experience to recognise the people who:

- tell you what they think you want to hear

- express a view thinking it might agree with your own

- are watching their backs

- try to hide things.

Be scrupulous in establishing the truth – ask questions until you are satisfied you have it right.

You are good at valuing if you can say that invariably you have good judgement and the converse is also true. Knowing the truth or reality can then be followed by deciding what to do.

Also, beware of inaccurate figures (even from accounts departments!), errors in facts, figures and assumptions and specious assurances – all must be tested for accuracy and 'truth'.

Decision-making and weighing up the options

It is invariably necessary to choose a particular course of action out of a range of possible 'options'. What is the best way of ensuring that your own selection process is a sound one? The basic point here is that you should never assume that there is only one option open to you. Consider a number of options (or as many sensible and pertinent ones as you can muster), many of which will be directly dictated or affected by the facts that you can establish. Gathering information also helps the collection of options, even considering options that you might think are closed to you (eg increasing price, scrapping low-profit items etc).

Selecting and working through a range of options means considering:

- Which are the possible ones?
- Which of those are feasible?
- How to reduce feasible options to two choices, the 'either/ or'?
- Which one to choose (or a mixture)?

- Whether any action is really necessary at all, now, later?
- Whether or not to keep options open, ie not to choose yet?

You should avoid any compulsion to take action through an option where no action would be better and you should avoid assuming that there are only two possibilities, until you have weighed up all the feasible ones you can in a reasonable timeframe.

Whilst considering the options beware false assumptions: test all for validity.

At the same time, it is essential to understand the other factors which can limit the range and choice of options or their applicability. Judgement (again beware false assumptions – including about these factors) is needed about:

1 **Time**

2 **Information**

3 **Resources**

4 **Knowledge.**

You have to know the real (not assumed) limits which the above factors can impose on the options available to you.

Generating options, particularly if, initially, there seems to be only one, will usually lead to better decision-making. This is where imagination, creative-thinking and intuition can help.

Considering fresh possibilities and suspending judgement whilst generating them (through brainstorming) can increase the range of options by avoiding negativity as in:

- 'It won't work'
- 'We do it this way'

- 'It can't be done'
- 'It failed before'.

In weighing the options you must refine your skills at considering the consequences, both the possible and the probable. This will lead to assessment of risk and reward, and both should be carefully calculated.

Can you accept the risk of failure – what is the worst that can happen if it fails and can I accept it?

Judgement then is used in selecting from the range of options which have been carefully weighed and assessed as to their probable outcomes.

It can be useful to clarify options or, in decision-making to seek to test by argument or discussion. If done in the right way clarity can be the result.

Disagreement should be encouraged, not for its own sake of course, but to stimulate ideas. Discussion prior to action should not be feared, but arguing badly (for example: being personal, over pernickety, seeking to procrastinate unnecessarily etc.) should be avoided.

Summary and six-month follow-up test

define objective

check information

implement

Do you have a
five point plan
approach to
decision-making
and problem-solving?

develop options

evaluate
and decide

Six-month follow up test

DO YOU USE

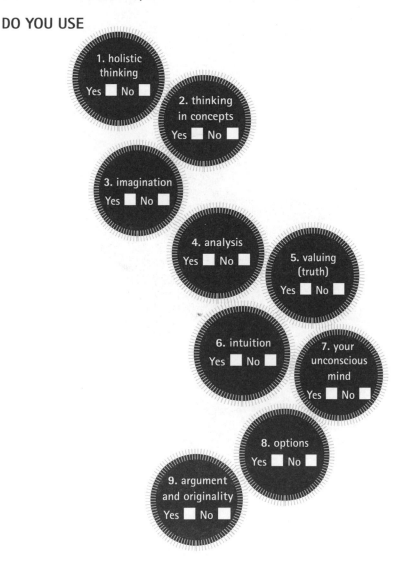

1. holistic thinking
Yes ☐ No ☐

2. thinking in concepts
Yes ☐ No ☐

3. imagination
Yes ☐ No ☐

4. analysis
Yes ☐ No ☐

5. valuing (truth)
Yes ☐ No ☐

6. intuition
Yes ☐ No ☐

7. your unconscious mind
Yes ☐ No ☐

8. options
Yes ☐ No ☐

9. argument and originality
Yes ☐ No ☐

IN YOUR PROBLEM-SOLVING?

04

CREATIVITY AND INNOVATION

04

Introduction

Really good managers (and all successful businesses have them) are capable of having, or recognising, good ideas and using them to make things happen in a new way: of translating ideas into useful, practicable and commercial products, services or systems.

Innovation (to bring in or introduce something new – a new idea, method or device) draws together new ideas *and* their implementation, whereas creativity is the having of new ideas which, in an organisation, are generated or spotted by individuals or teams.

It is important to:

- understand creativity and creative processes
- eliminate impediments to creativity
- widen the field of view
- build on ideas not merely criticise them
- tolerate doubts and uncertainties
- adopt a creative attitude in listening, observing and reading
- be confident in your own creative skills
- make time to think
- participate creatively as a leader, manager or member of a team
- use teams to innovate effectively
- manage innovation in your business.

This section of the book is divided into two parts: **creativity** (which looks at obstacles to creativity and ways to improve it personally and organisationally; and **innovation** (how best to manage creativity and innovate successfully in business).

Creativity

There are a number of *obstacles* which inhibit creativity. The seven main ones are:

1 Negativity
2 Fear of failure
3 Lack of quality thinking time
4 Over-conformance with rules and regulations
5 Making assumptions
6 Applying too much logic
7 Thinking you are not creative.

These obstacles can be seen in this identi-kit profile of the non-creative person; someone who is:

- not able to think positively about problems (and does not see them as opportunities)
- too busy or stressed to think objectively or at all
- very self-critical
- timid in putting forward a new idea (fearing ridicule)
- viewed as a conformist by friends/colleagues
- prone to apply logic as a first and last resort
- sceptical that many people are capable of being creative
- unable to think laterally
- uninspired even when confronted with a new idea.

On the other hand, creativity can be encouraged in people (including oneself) by exploring some of the qualities and characteristics of creative thinkers and the activities/steps that can be undertaken to improve the processes involved.

To be creative an individual should:

1 think beyond the invisible frameworks that surround problems/situations

2 recognise when assumptions are being made and challenge them

3 spot blinkered thinking and widen the field of vision (to draw on the experiences of other individuals/businesses)

4 develop/adapt ideas from more than one source

5 practice serendipity (finding valuable and agreeable things when not particularly seeking them) – having a wide attention span and range of interests is important

6 'transfer technology' from one field to another

7 be open/prepared to use chance or unpredictable things/events to advantage

8 explore thought processes and the key elements of the mind at work in analysing, valuing and synthesising

9 use his/her 'depth' mind (the unconscious mind) for example by sleeping on a problem to generate creative solutions to problems

10 note down thoughts/ideas that apparently drop into the mind unsolicited so that they are not forgotten

11 use analogy (to improve imaginative thinking) to find 'models' or solutions in 'nature', in existing products/services and/or in other organisations – not always reinventing the wheel

12 try, as appropriate, to sometimes make the strange familiar and the familiar strange to spark new ideas

13 make connections with points that are:
   • apparently irrelevant

- disguised/buried or not easily accessible
- outside own sphere of expertise
- lacking authority

14 suspend judgement to encourage the creative process and avoid premature criticism – analysis and criticism repress creativity)

15 know when to leave a problem (remaining aware but detached) for solutions to emerge – patience is important here as is the suspension of judgement

16 tolerate ambiguity and occasionally live with doubt and uncertainty

17 stimulate own curiosity (in everything including travel) and the skills of observation, listening, reading and recording.

Managers should remember that creativity should challenge the status quo to test continuously for improvements, because:

- a thing is not right because we do it
- a method is not good because we use it
- equipment is not the best because we own it

Creativity can be improved by remembering that the creative process has four main stages and each needs to be properly 'worked':

1 Preparation (information gathering, analysis and solution exploration)

2 Incubation (letting the mind work to continue the process)

3 Illumination (inspiration – which can come when the individual is not necessarily thinking about the problem but is in a relaxed frame of mind)

4 Verification (testing ideas, solutions, hunches, insights for applicability).

If ideas or solutions to problems are slow to come it sometimes can pay to leave matters alone for a while and reassess with:

- a new starting point;
- a different perspective;
- fresh motivation; and
- further consultation.

Innovation

It is worth identifying some of the key players who, if they were all present within an organisation would surely make it unbeatable.

Creative thinker: produces new and original ideas

Innovator: brings new products/services to the market or changes existing ones

Inventor: produces new and commercial ideas

Entrepreneur: conceives or receives ideas and translates them into business reality to exploit a market opportunity

Intrapreneur: responsible for innovation within an organisation

Champion: determination and commitment to implement an idea

Sponsor: backs an idea and helps remove obstacles

Successful businesses run on innovation and change, and effective innovation requires:

1 a blend of new ideas
2 the ability to get things done
3 sound commercial sense
4 customer focus; and
5 a conducive organisational climate.

Managers should be able to:

- manage for creativity
- provide an organisational environment in which innovation can thrive
- use a variety of techniques to stimulate ideas for products / services / systems and to generate ideas for bringing them to fruition.

To manage innovation (and draw 'greatness' out of people, it must be seen as a process with three phases:

1 **The generation of ideas** (from individuals and teams)
2 **The harvesting of ideas** (people evaluating ideas)
3 **The implementation of ideas** (teams developing and introducing ideas to the final, customer-satisfied stage).

In innovation it must be remembered that creative thinking makes it possible and teamwork makes it happen. Successful innovation requires an organisation and its key managers to be able to perform five essential functions, these are as follows:

1 Recruit/select creative people

For the appropriate jobs, of course, you will need creative people and their characteristics tend to be:

- high general intelligence
- strongly motivated
- stimulated by challenge
- vocational in attitude to work
- able to hold contradictory ideas together in creative tension
- curious with good listening and observing skills
- able to think for themselves; independent in thought
- neither an introvert nor an extrovert, but rather in the middle
- interested in many areas/things.

Creative individuals thrive if they are:

1 appreciated and receive recognition
2 given freedom to work in their area(s) of greatest interest
3 allowed contact with stimulating colleagues
4 given stimulating projects to work on
5 free to make mistakes.

2 Encouragement of creativity in teams

It is not always easy to manage the creative and innovative aspects of teamwork, where individuals ideally need to share the values, characteristics and interests of the other team members, to work with them in harmony and yet have something different to offer.

Belbin identified nine team-member roles, some of which are relatively self-explanatory:

1 **Plant** (solves difficult problems)

2 **Resource Manager**

3 **Co-ordinator**

4 **Shaper**

5 **Monitor/Evaluator**

6 **Team Worker** (co-operative, diplomatic)

7 **Implementer**

8 **Completer**

9 **Specialist.**

A good team will exhibit all of the above 'roles', not necessarily with nine different people, but with fewer team members adopting different roles to complete the task.

Encouraging creativity in teams (besides helping individuals to 'perform' the Belbin roles within a team) depends on a manager's skills at:

- using the different skills within the team (having first identified the attributes of each individual)

- ensuring conflicts of ideas are allowed to happen and are tolerated by all

- recognising particularly good contributions

- helping the team generate ideas (eg by brainstorming)

- creating an open environment where individuals can speak up honestly.

3 Team training

Self-evidently important (to improve team performance) is the development of team creativity by improving an individual's skills in effective thinking, communication and in his/her own particular area of expertise/specialism.

4 Communicating about innovation

Feedback can maintain interest levels and information about progress made can stimulate further activity and more progress. Good communication can help improve creativity and innovation and should:

- stress the importance of new ideas and show how business has improved because of their implementation
- indicate why ideas have been rejected/accepted
- give progress reports of ideas originated by individuals/teams
- recognise and reward appropriately for successful innovation.

5 Overcoming obstacles to innovation

Managers must ensure that creativity and innovation are not killed off by:

1 an initial response of outright condemnation, ridicule, rejection, damning criticism or faint praise
2 the vested interest of a particular person or department
3 too early an evaluation/judgement – sometimes suspending judgement early on can see an idea grow and reach a strong stage where it will work.

Managers who are creative and act in innovation-friendly ways have not only the usual leadership skills (of defining objectives, planning, controlling, supporting and reviewing in the areas of task, team and individual needs) but also are able to:

- accept risk
- work with half-formed ideas
- bend the rules
- respond quickly
- be enthusiastic (to motivate others).

Organisations and innovation

The business organisation itself has to provide an environment in which creativity and innovation can flourish and the five hallmarks of those organisations that actually are good at innovation (and not just paying lip service to it) are:

1 top level commitment
2 flexible in organisational structure
3 tolerant of failure (and not risk averse)
4 encouraging of team work and innovation
5 good at open and constructive communication.

'Managing innovation' is a 'challenge to management... especially top management and a touchstone of its competence'

Peter Drucker

Organisations need to work at the main ingredients for success at managing innovation and apply themselves to the five hallmarks listed above.

1 Top level commitment

This must be visible and audible and top management must ensure that blocks are removed and that inhibiting bureaucracy/individuals does not foul up the process. Chief executives and senior managers must value new ideas and innovation and participate actively to ensure that all know of their commitment to positive and useful change. Sometimes the need for short-term profits can dull the edge of creativity and innovation. Only top management can prevent this happening – taking the long not the short-term view.

2 Flexible in organisational structure

The antithesis of the innovative organisation is the bureaucratic one and Weber's characteristics of bureaucratic organisations are as follows:

- authority is impersonal and formal
- strong emphasis on functional specialisation
- a rule for every eventuality
- strong emphasis on hierarchy and status
- clearly laid down procedures (red tape)
- proliferation of paperwork
- security of employment and advancement by seniority.

At the opposite end of the scale would be the flexible organisation which is one:

- capable of responding to changing situations
- where barriers between staff in different areas are minimised
- with a flat rather than pyramid organisational structure
- where decision-making is pushed downwards to where the organisation meets its customers/suppliers
- with entrepreneurial flair present at all levels
- which can develop and test more than one solution to problems encountered
- with efficient rather than stifling monitoring systems
- which has enough 'discipline' to get things done
- which balances freedom and order.

3 Tolerant of failure

Innovation and risk go hand in hand. Management which goes into critical overdrive when mistakes occur (rather than analysing them to learn from the failures) smothers creativity and innovation. Risks can yield failure, but not taking risks can spell total disaster and an end to profits and growth.

Unless failure results from negligence, recklessness or complete incompetence, managers should not seek out scapegoats or exact revenge. Profits are the reward for taking risks and innovative organisations learn to live with risk.

4 Encouraging teamwork and innovation

In innovation it can be said that none of us is as good as all of us. Teamwork and innovation are better in organisations where:

- the climate is open
- participation is encouraged
- facts and information are readily available
- change is managed positively
- resources are provided for training and development
- rules are at a minimum (with policies and guidelines instead)
- internal communications are good and more by mouth than memo
- respect is given to all colleagues (but not on demand by management – it has to be earned)
- managers are themselves highly motivated
- teamwork often transcends departmental boundaries.

5 Good at open and constructive communication

Communication should be good laterally **and** vertically (and flatter organisations should – in theory, at least – encourage good lateral communication). Managers should ensure a good flow of information – ideas can emerge as a result. Cross-fertilisation can create more (and better) ideas, particularly where departmental or divisional boundaries are crossed.

Checklist for the innovative organisation

	YES	NO
Is the top management team committed to innovation?	☐	☐
Does the organisation express clearly its vision (which should include an emphasis on innovation)?	☐	☐
Is the Chief Executive openly enthusiastic for change?	☐	☐
Are mutual stimulation, feedback and constructive criticism all at high levels of activity?	☐	☐
Is the organisation good at team work including the use of project teams?	☐	☐
Are mistakes and failures accepted as part of risk-taking?	☐	☐
Do creative people join and stay with the organisation?	☐	☐
Is innovation rewarded (financially or by promotion or both)?	☐	☐
Are lateral communications good?	☐	☐
Can ideas be exchanged informally and are opportunities provided to do this?	☐	☐
Does the organisation raise excuses not to innovate?	☐	☐
Are resources given to new ideas?	☐	☐
Is the structure flexible?	☐	☐
Is decision-making pushed down to the lowest level at which decisions could be taken?	☐	☐
Do all staff see themselves as part of the creative and innovative processes?	☐	☐
Does the organisation take a long-term view of the benefits of innovation?	☐	☐
Is innovation part of the organisation's vision and strategy?	☐	☐
Is it fun to work in your organisation?	☐	☐

Scoring: Yes/No – More than 9 No scores – not good!

The Generation of ideas

It is interesting to note that organisations can get ideas from, amongst other sources:

- R&D (internal or external)
- Staff
- Competitors
- Suppliers
- Customers
- Quality circles.

One survey demonstrated that SMEs (small and medium-sized enterprises) can get ideas from, in order of importance:

1 Staff
2 Customers
3 Market and competition
4 Board and Planning Group
5 Sales Department
6 Suppliers
7 Managing Director
8 Consultants
9 Shows and exhibitions.

Ideas have to be sieved – by individuals or by teams – to choose and refine the selected ideas to then develop them and take them to market.

Suggestion schemes can work provided people on all sides know that:

- all ideas from everyone will be listened to
- every idea deserves thanks
- some ideas will not work
- a forum for ideas assists the process of innovation.

Recognition (and selection) of ideas to be pursued should be on the basis that the idea can show:

- originality of thought
- ultimate benefit to the customer
- business potential
- quality improvement
- cost savings
- viability in implementation.

In sieving ideas, three questions should be asked (as Henry Ford did):

1 Is it needed?
2 Is it practical?
3 Is it commercial?

Project teams can drive an idea forward to its successful innovation by remembering three strategies which work and they are to:

1 **recruit a senior sponsor**
2 **run a pilot project or experiment**
3 **present innovation as a gradual/incremental development.**

Brainstorming (getting a large number of ideas from a group in a short time) can produce ideas (which then have to be sieved and tested) and Alexander Osborn's rules are hard to beat:

- **Suspend judgement – no criticism or evaluation**
- **Free-wheel – anything goes, the wilder the better**
- **Quantity – the more ideas the merrier**
- **Combine and improve – link ideas, improve suggested ones.**

**In leading a brainstorming session
the four main steps are:**

1	**Introduce**	aim of session and remind people of Osborn's rules
2	**Warm-up**	if necessary do a practice exercise (eg 20 uses for a hammer)
3	**State the problem**	not too detailed
4	**Guide**	time to think
		generation of ideas
		no judgement/criticism/evaluation!
		clarify
		maintain free-flow of ideas.

In leading a session which is 'sticky' and short of ideas to start with, ask 'what if' questions to stimulate thought.

Brainstorming sessions should always be followed up, perhaps in smaller groups and ideas should then be evaluated by:

- deciding the selection criteria
- selecting obvious winning ideas
- eliminating the unworkable ideas

- sifting ideas into groupings and selecting the best in each
- applying the selection criteria to obvious winners and 'best of' the various groups
- testing the selections by 'reverse brainstorming' (ie in how many ways can this idea fail?)
- informing the participants of further developments.

Successful brainstorming by managers can be achieved by asking yourself these questions:

1 Do you use it whenever appropriate?
2 Does it work? If not, are you leading it effectively?
3 Is there a better person than you to lead a session?
4 Can you point to where brainstorming sessions have improved creative thinking in your organisation?
5 Do you and your managers have a list of problems that could benefit from brainstorming?
6 Do you use teams sufficiently to work on problems?

In taking good ideas to market there are a number of questions you should apply to your organisation.

Checklist for the generation of ideas

	YES	NO
Is there an internal market for innovative ideas?	☐	☐
Do teams allocate time to consider ideas?	☐	☐
Do you and your teams spend time away from the office to review performance and plans?	☐	☐
Are customers/suppliers involved in innovation in your business?	☐	☐
Do you have successfully innovative teams and/or individuals and can you identify reasons for their success?	☐	☐
Do you have a suggestion scheme that works?	☐	☐
Are new ideas properly rewarded?	☐	☐
Do you help ensure ideas are not lost through poor presentation?	☐	☐
Do you know of an alternative route to profitability and growth other than through innovation?	☐	☐

Characteristics of innovators

As a manager you should possess these skills/qualities:

1 **A clear vision of the required results**

2 **The ability to:**

- define objectives and benefits of ideas/projects
- present the case for it powerfully
- win support from superiors, colleagues and subordinates

- motivate people into action so that they contribute to project success
- influence others to support the project
- deal with criticism, interference, lack of enthusiasm, disputes and lateness

3   Courage – to take risks (and endure any setbacks)
4   Willpower to maintain momentum
5   Fairness – to ensure all who participate are recognised/ rewarded.

Summary and six-month follow-up test

The seven habits of successful creative thinkers are:

1   Thinking outside the apparent confines of the problem/ situation
2   Welcoming chance intrusions
3   Listening to your depth mind (the unconscious mind)
4   Suspending judgement
5   Using the stepping stones of analogy
6   Tolerating ambiguity
7   Banking all ideas from all sources.

Innovation needs the generation, harvesting and implementation of ideas. Managers good at innovation accept risk, are flexible and are motivated to take ideas through to completion.

Six-month follow-up test

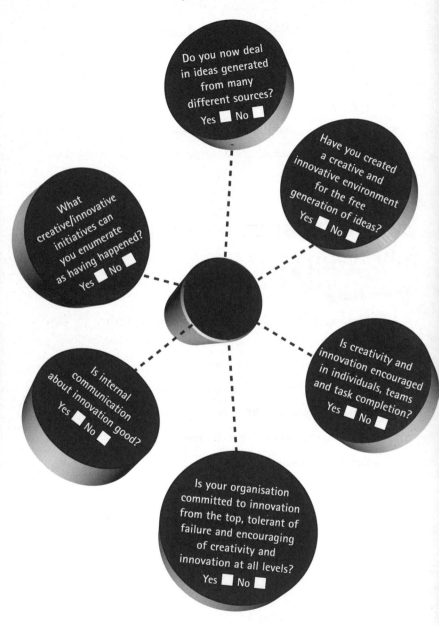

Do you now deal in ideas generated from many different sources?
Yes ☐ No ☐

Have you created a creative and innovative environment for the free generation of ideas?
Yes ☐ No ☐

What creative/innovative initiatives can you enumerate as having happened?
Yes ☐ No ☐

Is creativity and innovation encouraged in individuals, teams and task completion?
Yes ☐ No ☐

Is internal communication about innovation good?
Yes ☐ No ☐

Is your organisation committed to innovation from the top, tolerant of failure and encouraging of creativity and innovation at all levels?
Yes ☐ No ☐

05

PART 1:
PERSONAL REMINDERS
AND THOUGHTS WORTH
THINKING

05

Introduction

This section will look at the main areas covered in the first part of the book and will offer:

i) personal reminders – points to bear in mind in pursuing your personal development, particularly to ensure that you achieve a balance between personal and business success; and

ii) thoughts worth thinking – quotes from various sources which shed light on the management topics covered in Part 1, which you might find helpful or inspiring.

Time management

i) Personal reminders

The Adair ten principles of time management		
	1	Develop a personal sense of time
	2	Identify long-term goals
	3	Make medium-term plans
	4	Plan the day
	5	Make the **best** use of your **best** time
	6	Organise office work
	7	Manage meetings
	8	Delegate effectively
	9	Make use of committed time
	10	Manage your health

ii) Thoughts worth thinking

'Nothing really belongs to us but time, which even he has who has nothing else.'

Baltasar Gracian

'Time wasted is existence, used is life.'

Edward Young

'I know what time is until someone asks me.'

St Augustine

'Become time obsessed. Virtually all staff processes can be shortened by between 50 per cent and 99 per cent. Directly tie performance evaluations to speed. Remember: 'What gets measured gets done'. Make speed pay!'

Tom Peters

*'Never put off until tomorrow
what you can do today.'*

Lord Chesterfield

A personal sense of time

*There is nothing which I wish more that you should know,
and which fewer people do know, than the true use and
value of time. It is in everybody's mouth, but in few
people's practice. Every fool, who scatters away his whole
time in nothing, utters, however, some trite commonplace
sentence, of which there are millions, to prove, at once, the
value and the fleetness of time. The sundials all over
Europe have some ingenious inscriptions to that effect; so
that nobody squanders away their time without hearing
and seeing daily how necessary it is to employ well, and
how irrecoverable it is if lost.*

Lord Chesterfield • Letters to His Son

*'Never leave till tomorrow
which you can do today.'*

Benjamin Franklin

'It's not enough to be busy.
The question is:
'What are you busy about?'

Henry Thoreau

'Time is the scarcest resource
and unless it is managed
nothing else can be managed.'

Peter Drucker

'If I were asked what one thing an executive could do that
would really, and quickly, make him more effective, make
him achieve more and make him enjoy what he's doing, I'd
say: Make sure you know where your time goes. Don't
depend on memory, it's treacherous.'

Peter Drucker

'An inch of **gold** cannot buy an **inch** of time.'

Chinese proverb

He who every morning plans the transactions of the day and follows out that plan carries a thread that will guide him through the labyrinth of the most busy life. The orderly arrangement of his time is like a ray of light which darts itself through all his occupations. But where no plan is laid, where the disposal of time is surrendered merely to the chance of incidents, chaos will soon reign.

Victor Hugo

'It is not possible to hold the day. It is possible to lose it.'

Lines on a Sundial, 1695

'Do you love life? Then do not squander time, for that is the stuff life is made of.'

Benjamin Franklin

'In the morning when I get up, the first thing I do is think to myself: what am I going to do today? So many things: I count them, think about them and to each I assign its time… I'd rather lose sleep than time, in the sense of the proper time for doing what has to be done… to watch the time, to devote oneself to business and never lose an hour of time… He who knows how not to waste time can do just about anything; and he who knows how to make use of time will be Lord of whatever he wants.'

Leon Alberti

'Doing the day's business in the day.'

Wellington

'He slept beneath the moon,
He basked beneath the sun
He lived a life of going-to-do
And died with nothing done.'

Epitaph of James Albery (1838-1889)

**'What wears one out is not what one does
but what one doesn't do.'**

Archbishop William Temple

'It is essential for the Cabinet to move on,
leaving in its wake a trail of clear, crisp,
uncompromising decisions. That is what
government is about. And the challenge to
democracy is to get it done quickly.'

Clement Attlee

'Time is a dressmaker specialising in alterations.'

Faith Baldwin

'Go, Sir, gallop, and don't forget that the world was made in six days. You can ask me for anything you like, except time.'

Napoleon

'Procrastination is the thief of time.'

English Proverb

'Time wastes our bodies and our wits,
But we waste Time, so we are quits.'

Lines on a Sundial, 1746

'Remember that time is **money!**'

Benjamin Franklin

Controlling the Cabinet

The Cabinet usually meets once a week. That should be enough for regular meetings, and should be if they grasp from the start what they are there for. They should be back at their work as soon as possible, and a Prime Minister should put as little as possible in their way. We started sharp at 11, and rose in time for lunch. Even in a crisis, another couple of meetings should be enough in the same week: if there is a crisis, the less talk the better.

The Prime Minister shouldn't speak too much himself in Cabinet. He should start the show or ask somebody else to do so, and then intervene only to bring out the more modest chaps who, despite their seniority, might say nothing if not asked. And the Prime Minister must sum up. Experienced Labour leaders should be pretty good at this; they have spent years attending debates at meetings of the Parliamentary Party and the

National Executive, and have to sum those up. That takes some doing — good training for the Cabinet.

Particularly when a non-Cabinet Minster is asked to attend, especially if it is his first time, the Prime Minister may have to be cruel. The visitor may want to show how good he is, and go on too long. A good thing is to take no chance and ask him to send the Cabinet a paper in advance. The Prime Minister can then say, 'A very clear statement, Minister. Do you need to add anything?' in a firm tone of voice obviously expecting the answer, 'No'. If somebody else looks like making a speech, it is sound to nip in with 'Has anybody any objection?' If somebody starts to ramble, a quick, 'Are you objecting? You're not? Right. Next business,' and the Cabinet can move on.

Clement Attlee

'You cannot kill time without injuring eternity.'

Henry Thoreau

'The unexpected caller, pleasant or unpleasant, disturbs the pattern of your time. That time is already allotted. Hence there is no time to help somebody who needs it, to relax and enjoy a chat on the street.

Enjoyment is always for **after** when all the jobs with deadlines, the letters to be answered, the calls to be made, have been despatched. But that time never comes. There are always more letters, deadlines, jobs and so life gets postponed until an indefinite after – until it is too late. A hamster on a treadmill has about the same sort of freedom.'

Dr Jonathon Steinberg

The daily miracle

Philosophers have explained space. They have not explained time. It is the inexplicable raw material of everything. With it, all is possible; without it, nothing. The supply of time is truly a daily miracle, an affair genuinely astonishing when one examines it. You wake up in the morning, and lo! your purse is magically filled with twenty-four hours of the unmanufactured tissue of the universe of your life! It is yours. It is the most precious of possessions. A highly singular commodity, showered upon you in a manner as singular as the commodity itself!

For remark! No one can take it from you. It is unstealable. And no one receives either more or less than you receive.

Talk about an ideal democracy! In the realm of time there is no aristocracy of wealth, and no aristocracy of intellect. Genius is never rewarded by even an extra hour a day. And there is no punishment. Waste your infinitely precious commodity as much as you will, and the supply will never be withheld from you. No mysterious power will say: 'This man is a fool, if not a knave. He does not deserve time; he shall be cut off at the meter'. It is more certain than government bonds, and payment of income is not affected by Sundays. Moreover, you cannot draw on the future. Impossible to get into debt! You can only waste the passing moment.

You cannot waste tomorrow; it is kept for you. You cannot waste the next hour; it is kept for you.

I said the affair was a miracle. Is it not?

You have to live on this twenty-four hours of daily time. Out of it you have to spin health, pleasure, money, content, respect, and the evolution of your immortal soul. Its right use, its most effective use, is a matter of the highest urgency and of the most thrilling actuality. All depends on that.

Arnold Bennett

'Lord,

there's never enough time for everything.

Help

me to do a little less a little better.'

Anon

On turning over a new leaf

The most important preliminary to the task of arranging one's life so that one may live fully and comfortably within one's daily budget of twenty-four hours is the calm realisation of the extreme difficulty of the task, of the sacrifices and the endless effort which it demands. I cannot too strongly insist on this.

If you imagine that you will be able to achieve your ideal by ingeniously planning out a timetable with a pen on a piece of paper, you had better give up hope at once. If you are not prepared for discouragements and disillusions, if you will not be content with a small result for a big effort, then do not begin. Lie down again and resume the uneasy doze which you call your existence.

It is very sad, is it not, very depressing and sombre? And yet I think it is rather fine, too, this necessity for the tense bracing of the will before anything worth doing can be done. I rather like it myself. I feel it to be the chief thing that differentiates me from the cat by the fire.

'Well', you say, 'assume that I am braced for battle. Assume that I have carefully weighed and comprehended your ponderous remarks; how do I begin?' Dear sir, you simply begin. There is no magic method of beginning. If a man standing on the edge of a swimming-bath and wanting to jump into the cold water should ask you, 'How do I begin to jump?' you would merely reply, 'Just jump. Take hold of your nerves, and jump.'

As I have previously said, the chief beauty about the constant supply of time is that you cannot waste it in advance. The next year, the next day, the next hour are lying ready for you, as perfect, as unspoilt, as if you had never wasted or misapplied a single moment in all your career. Which fact is very gratifying and reassuring. You can turn over a new leaf every hour if you choose. Therefore no object is served in waiting till next week, or even until tomorrow. You may fancy that the water will be warmer next week. It won't. It will be colder.

Arnold Bennett

Setting and achieving goals and objectives

i) Personal reminders

1 Define the purpose

2 Define your strategic aims

3 Identify long-term goals

4 Make middle-term plans (for middle-term goal achievement on the road to long-term goal achievement)

5 Set goals and objectives

6 Know your values (and review them)

7 Budget your time

8   Review goals and objectives and progress in their achievement,
    regularly
9   Adjust plans and activities to achieve goals set/adjusted.

ii) Thoughts worth thinking

'If you do not know
where you are going,
you can take any road.'

Anon

'It is

wise

to look

ahead,

but foolish

to look

further than

you can

see.'

Winston Churchill

'If you want to make anything happen you must make time and space for it.'

Anon

a zest for living

If you want to enjoy one of the greatest luxuries in life, the luxury of having enough time, time to rest, time to think things through, time to get things done and know you have done them to the best of your ability, remember there is only one way. Take enough time to think and plan things in the order of their importance. Your life will take a new zest, you will add years to your life, and more life to your years. Let all your things have their places. Let each part of your business have its time.

Benjamin Franklin

Decision-making and problem-solving

i) Personal reminders

- The ability to take decisions – the most valuable skill in management
- The five point plan:
  - Define objective
  - Check information
  - Develop options
  - Evaluate and decide
  - Implement.
- Use: analysis, holistic thinking, thinking in concepts, imagination, valuing (truth), intuition, your unconscious mind, options, argument, originality
- Know your mind and develop your thinking skills.

ii) Thoughts worth thinking

'If I have any advice to pass on it is this: if one wants to be successful one must think until it hurts... Believe me, this is hard work and, from my close observation, I can say that there are few people indeed who are prepared to perform this arduous and tiring work.'

Roy Thomson

'When I go into any business deal my chief thoughts are on how I'm going to save myself if things go wrong.'

Paul Gretty

'The final act of business judgement is intuitive.'

Anon

'I take it we are all in complete agreement on the decision here ... then, I propose we postpone further discussion of this matter until our next meeting to give ourselves time to develop disagreement and perhaps gain some understanding of what the decision is about!'

Alfred P Sloan

'Rightly to be great is not to stir without great argument.'

Shakespeare

'Men sleep well in the Inn of Decision.'

Old Arab Proverb

Creativity and innovation

i) Personal reminders

The seven habits of successful creative thinkers are:

1 Thinking outside the apparent confines of the problem/situation

2 Welcoming chance intrusions

3 Listening to your depth mind (the unconscious mind)

4 Suspending judgement

5 Using the stepping stones of analogy

6 Tolerating ambiguity

7 Banking ideas from all sources

Innovation needs the generation, harvesting and implementation of ideas and managers good at innovation accept risk, are flexible and are motivated to take ideas through to completion.

iii) Thoughts worth thinking

'He that will not apply new remedies must accept new evils – for time is the greatest innovator.'

Francis Bacon

'DARING IDEAS ARE LIKE CHESSMEN

moved forward.

THEY MAY BE BEATEN, BUT THEY MAY START A

winning game.'

Goethe

'Experience has shown and a true philosophy will always show that a vast, perhaps the larger, portion of truth arises from the seemingly irrelevant.'

Edgar Allan Poe

'Discovery consists of seeing what everyone has seen and thinking what nobody has thought.'

Anon

'Thinking will always give you a reward, though not always what you expected.'

Roy Thomson

'In the case of the creative mind, it seems to me it is as if the intellect has withdrawn its guards from the gates. Ideas rush in pell mell and only then does it review and *examine the multitude. You worthy critics, or whatever you may call yourselves, are ashamed or afraid of the momentary and passing madness found in all real creators… Hence your complaints of unfruitfulness – you reject too soon and discriminate too severely.'*

Johann Schiller

'While the fisher sleeps, the net takes the fish.'

Old Greek Proverb

'A person would do well to carry a pencil in their pocket and write down the thoughts of the moment. Those that come unsought are commonly the most valuable and should be secured because they seldom return.'

Francis Bacon

'A new idea is delicate.
It can be killed by a sneer or a yawn:
it can be stabbed to death by a quip
and worried to death by a frown on
the right man's brow.'

Charles Brower

'God

HIDES THINGS FROM US BY PUTTING THEM CLOSE TO US.'

Old saying

'Criticism often takes
from the tree caterpillars
and blossoms together.'

Old saying

*Don't be afraid of taking a big step – you
cannot cross a chasm in two steps.'*

David Lloyd-George

'The

mind

connects

things in

unbelievable

ways.'

George Benjamin

'A man without patience is a lamp without oil.'

Andres Segovia

'One should never impose one's views on a problem; one should rather study it and in time a solution will reveal itself.'

Albert Einstein

'I invented nothing: I rediscover.'   Rodin

'As soon as a thought darts, I write it down.'

Thomas Hobbes

'My chief job is to constantly stir or rekindle the

curiosity of people that gets driven out by

bureaucracy and formal schooling systems.'

Akio Morito

'There is a great deal of unmapped country within us.'

English Proverb

'Curiosity is one of the permanent and certain

characteristics of a vigorous intellect.'   Samuel Johnson

'One doesn't discover new lands without consenting to lose sight of the shore for a very long time.'   Andre Gide

'I question.'

Leonardo da Vinci's motto

'The disease of not listening.'

Shakespeare

'Many ideas grow better when transplanted into another mind than in the one where they sprung up.'

Oliver Wendell Holmes Jr

'Man never rises to great truths without enthusiasm.'

Vauvenargues

'The typical eye

sees the ten per cent bad

of an idea and overlooks

the ninety per cent good.'

90%

Charles F Kettering

'More creativity is the only way to make tomorrow better than today.'

Anon

The creative art thrives in an environment of mutual stimulation, feedback and constructive criticism – in a community of creativity.'

William T Brady

'Chance favours only the prepared mind.'

Louis Pasteur

'Between the idea

And the reality...

Falls the Shadow.'

T S Eliot

'An established company which in an age demanding innovation is not capable of innovation is doomed to decline and extinction. And a management which in such a period does not know how to manage innovation is incompetent and unequal to its task. Managing innovation will increasingly become a challenge to management, and especially top management, and a touchstone of its competence.'

Peter Drucker

The most important of my discoveries have been suggested to me by failures.'

Sir Humphrey Davy

Innovation is a gamble!

Sydney Brenner

'Without real commitment from the top,

real innovation will be defeated again and again

by the policies, procedures and rituals

of almost any large organisation.'

Anon

There are costs and risks to a programme of action.
But they are far less than the long-range risks and costs
of comfortable inaction.' John F Kennedy

'The way to be safe is never to be secure.'

'Experience is the name we give our mistakes.'

Oscar Wilde

'What is honoured in a country will be cultivated there.'

Plato

'There is a natural opposition among men to anything they have not thought of themselves.'

Barnes Wallis

'People support what they help to create.'

Anon

'What's the secret of entrepreneurial success? It's knowing when to use OPB (Other People's Brains) and OPM (Other People's Money).'

J B Fugua

'Changing things is central to leadership. Changing them before anyone else is creativeness.'

Anon

'He that wrestles with us strengthens our nerves and sharpens our skill. Our antagonist is our helper.'

Edmund Burke

'He who dares **nothing,** for **need hope nothing.'**

English Proverb

PART TWO:
MANAGING OTHERS

two

06

LEADERSHIP AND TEAMBUILDING

06

This chapter of the book is divided into two parts:

A. Leadership and **B. Teambuilding.**

A. Leadership

A survey of successful chief executives on the attributes most valuable at top levels of management indicated the following in order of rating:

1   *Ability to take decisions*

2   *LEADERSHIP*

3   *Integrity*

4   *Enthusiasm*

5   *Imagination*

6   *Willingness to work hard*

7   *Analytical ability*

8   *Understanding of others*

9   *Ability to spot opportunities*

10 *Ability to meet unpleasant situations*

11 *Ability to adapt quickly to change*

12 *Willingness to take risks*

13 *Enterprise*

14 *Capacity to speak lucidly*

15 *Astuteness*

16 *Ability to administer efficiently*

17 *Open-mindedness*

18 *Ability to 'stick to it'*

19 *Willingness to work long hours*

20 *Ambition*

21 *Single-mindedness*

22 *Capacity for lucid writing*

23 *Curiosity*

24 *Skill with numbers*

25 *Capacity for abstract thought.*

There is (has and probably always will be) a debate about the differences and overlaps of leadership and management. Current opinion is that they are different concepts but they overlap considerably.

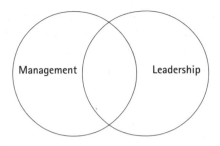

Perhaps management has the overtone of carrying out objectives laid down by someone else. It is certainly true that a well-managed business, in the sense of having perfect organisation, still needs that extra something.

Leadership has five distinctive nuances not found in management.

A leader must:

1 **Give direction**
2 **Provide inspiration**
3 **Build teams**
4 **Set an example**
5 **Be accepted.**

Henri Fayol (in 1916) divided the activities of an industrial company into six main groups:

1 **Technical:** production, manufacture, adaptation
2 **Commercial:** buying, selling, exchange
3 **Financial:** search for and optimum use of capital
4 **Security:** protection of property and people
5 **Accounting:** stocktaking, balance sheet, costs, statistics
6 **Administration:** forecasting and planning, organising, commanding, co-ordinating and controlling.

Good administration is the hallmark of good management and the proper and efficient use of resources. Managers become leaders when their personality and character, their knowledge and functional skills of leadership are recognised and accepted by the others involved.

Leadership can be 'specific to the particular situation' and its 'authority' can derive from:

1 position (as in job title, rank or appointment),

2 personality (as in natural qualities of influence) and

3 knowledge (as in technical professional skills).

Fayol listed these following qualities as being needed by a person in 'command'. A person in command should:

- have a thorough knowledge of employees

- eliminate the incompetent

- be well versed in the agreements binding the business and its employees

- set a good example

- conduct periodic audits of the organisation and use summarised charts to further this review

- bring together the chief assistants by means of conferences at which unity of direction and focusing of effort are provided for

- not become engrossed in detail

- aim at making unity, energy, initiative and loyalty prevail among all employees.

Qualities of leadership

A leader is the kind of person (with leadership qualities) who has the appropriate knowledge and skill to lead a group to achieve its ends willingly. This section looks at the qualities and functions of leadership.

Personality and character cannot be left out of leadership. There are certain generic leadership traits and seven important ones are:

Seven qualities of leadership

1 **Enthusiasm**: try naming a leader without it!

2 **Integrity:** meaning both personal wholeness and sticking to values outside yourself, primarily goodness and truth – this quality makes people trust a leader

3 **Toughness:** demanding, with high standards, resilient, tenacious and with the aim of being respected (not necessarily popular)

4 **Fairness:** impartial, rewarding / penalising performance without 'favourites', treating individuals differently but equally

5 **Warmth:** the heart as well as the mind being engaged, loving what is being done and caring for people – cold fish do not make good leaders

6 **Humility:** the opposite of arrogance, being a listener and without an overwhelming ego

7 **Confidence:** not over-confidence (which leads to arrogance), but with self-confidence which people know whether you have or have not got it

In testing whether or not you have the basic qualities of leadership, you should ask yourself these questions.

A checklist to test if you have the basic qualities of leadership

	YES	NO
Do I possess the above mentioned seven qualities? (This 'test' will subsequently reveal whether or not you really do!)	☐	☐
Have I demonstrated that I am a responsible person?	☐	☐
Do I like the responsibility and the rewards of leadership?	☐	☐
Am I well-known for my enthusiasm at work?	☐	☐
Have I ever been described as having integrity?	☐	☐
Can I show that people think of me as a warm person?	☐	☐
Am I an active and socially participative person?	☐	☐
Do I have the self-confidence to take criticism, indifference and/or unpopularity from others?	☐	☐
Can I control my emotions and moods or do I let them control me?	☐	☐
Have I been dishonest or less than straight with people who work for me over the past six months?	☐	☐
Am I very introvert, very extrovert (or am I an ambivert – mixture of both – as leaders should be)?	☐	☐

If leadership depends on the situation, you need to ask yourself, whatever your qualities, whether you are right for the situation:

A checklist to test if your qualities are right for the situation	YES	NO
Are your interests, aptitudes and temperament suited to your current field of work?		
If not, can you identify one that would better suit you where you would emerge as a leader?		
Do you have the 'authority of knowledge' in your current field (and have you acquired all the necessary professional and specialist skills through training that you could have done at this point in your career?)		
Are you experienced in more than one field/industry/function?		
Are you interested in fields adjacent and relevant to your own?		
Do you read situations well and are you flexible in your approach to changes within your field?		

Functions of leadership

In leadership, there are always three elements or variables:

1 **The leader:**   qualities of personality and character

2 **The situation:** partly constant, partly varying

3 **The group:**   the followers: their needs and values

This section of the book looks at leadership functions in relation to the needs of work groups. These needs can be seen as three overlapping needs:

1 **Task need:**   to achieve the common task

2 **Team maintenance needs:**   to be held together or to maintain themselves as a team

3 **Individual needs:**   the needs which individuals bring with them into the group.

These three needs (the task, team and individual) are the watchwords of leadership and people expect their leaders to:

- help them achieve the common task
- build the synergy of teamwork and
- respond to individuals and meet their needs.

The **task** needs work groups or organisations to come into being because the task needs doing and cannot be done by one person alone. The task has needs because pressure is built up to accomplish it to avoid frustration in the people involved if they are prevented from completing it.

The **team maintenance** needs are present because the creation, promotion and retention of group/organisational cohesiveness is essential on the 'united we stand, divided we fall' principle.

The **individual** needs are the physical ones (salary) and the psychological ones of:

- recognition
- a sense of doing something worthwhile

- status
- the deeper need to give and to receive from other people in a working situation.

The Task, Team and Individual needs overlap:

This overlapping is evident in that:

- achieving the task builds the team and satisfies the individuals
- if team maintenance fails (the team lacks cohesiveness) performance on the task is impaired and individual satisfaction is reduced
- if individual needs are not met the team will lack cohesiveness and performance of the task will be impaired.

Leadership exists at different levels:

Team leadership: of teams of about 5 to 20 people

Operational leadership: a significant must in a business or organisation comprising a number of teams whose leaders report to you

Strategic leadership: a whole business or organisation, with overall accountability for the levels of leadership below you.

At whatever level of leadership, Task, Team and Individual needs must be constantly thought about. To achieve the common task, maintain teamwork and satisfy the individuals, certain functions have to be performed. A function is what leaders *do* as opposed to a quality which is an aspect of what they *are*.

These functions (the functional approach to leadership, also called action-centred leadership) are:

- Defining the task
- Planning
- Briefing
- Controlling
- Evaluating
- Motivating
- Organising
- Providing an example.

Leadership functions in relation to Task, Team and Individual can be represented by this diagram:

Leadership functions

THE ROLE | FUNCTIONS

Achieving the TASK

Building and maintaining the TEAM

Developing the INDIVIDUAL

... Defining the task
... Planning
... Briefing
... Controlling
... Evaluating
... Motivating
... Organising
... Providing an example

These leadership functions need to be handled with excellence and this is achieved by performing those functions with increasing skill.

Before examining the skills of leadership, it is worth seeing where certain **qualities** of leadership can be viewed as having functional value. These can be examined as leadership characteristics.

Leadership characteristics

The Need	Quality	Functional Value
Task	Initiative	gets the group moving
	Perseverance	prevents the group giving up
	Efficiency	work done well knowing costs (energy, time and money)
	Honesty	establishing facts
	Self-confidence	facing facts
	Industry	steady application pays dividends
	Audacity	when not to be restrained by rules or convention
	Humility	facing up to mistakes and not blaming others
Team	Integrity	integrating the team and creating trust
	Humour	relieving tension and maintaining a sense of proportion
	Audacity	inspire through originality or verve
	Self-confidence	trusted by others

	Justice	fair dealing builds group discipline
	Honesty	wins respect
	Humility	not selfish, shares praise, not arrogant and divisive
Individual	Tact	sensitive in dealing with people
	Compassion	sympathetic awareness and help
	Consistency	people know where they stand
	Humility	recognises qualities/ abilities and gives credit
	Honesty	wins individual respect
	Justice	fair-dealing encourages individuals

Leaders need to exhibit the following attributes/qualities/characteristics in exercising the functions:

1 Group influence: to generate a group willingness to achieve a desired goal/objective

2 Command: to decide upon a course of action as quickly as the situation demands and to carry it through with a firmness and strength of purpose

3 Coolness: to remain cool or unperturbed under testing or trying conditions

4 **Judgement:**	ability to arrange available resources and information in a systematic and commonsense way to produce effective results
5 **Application/ responsibility:**	to demonstrate sustained effort combined with a degree of dependability in order to complete a task or achieve an objective.

Leadership skills

Having identified the main functions or principles of leadership, there are skills in providing those functions in different situations and managers need to develop their abilities to bring those skills to bear in increasing levels of excellence.

The eight functions (defining the task, planning, briefing, controlling, evaluating, motivating, organising and setting an example) will now be examined.

1 Defining the task

A task is something that needs to be done. People in organisations and teams need to have this distilled into an objective which is:

- clear
- concrete
- time-limited
- realistic
- challenging
- capable of evaluation.

There are five tests to apply to the defining of a task and they are:

i) Do you have a clear idea of the objectives of your group now and for the next few years/months which have been agreed with your boss?

ii) Do you understand the overall aims and purpose of the organisation?

iii) Can you set your group's objectives into the context of those larger intentions?

iv) Is your present main objective specific, defined in terms of time and as concrete/tangible as you can make it?

v) Will the team know for itself if it succeeds or fails and does it get speedy feedback of results?

In defining the task/communicating the objective you need to have the following abilities:

- To tell the group the objective you have been given
  BEWARE: *not understanding it yourself can lead to lack of clarity.*

- To tell the group what to do **and** why
  BEWARE: *giving the reason in terms of a past event rather than future.*

- To break down aims into objectives for other groups
  BEWARE: *not making them specific enough or not making sure there are enough objectives which add up to complete the aim.*

- To agree the objective
  BEWARE: *taking things for granted and not fixing on the objective.*

- To relate the aim to the purpose (to answer what and why questions)

> **BEWARE:** *confusing your division's aim with the purpose of the organisation.*

- To define the purpose and check that the aims relate to it and to each other
  BEWARE: *not doing it often enough.*

- To redefine the purpose to generalise it and create more aims and objectives
  BEWARE: *causing confusion by doing it too often or not knowing that it has to be done.*

- To communicate purpose to employees
  BEWARE: *using the wrong language, by-passing leaders below you or relying on others doing it for you.*

In defining the task, it needs to be broken down into objectives, aims and purpose so that it can be communicated with clarity. The end of the task should also be defined when the need arises and all should be aware of what the success criteria will be.

2 Planning

This key activity for any team or organisation requires a search for alternatives and that is best done with others in an open-minded, encouraging and creative way. Foreseeable contingencies should always be planned for.

Planning requires that the what, why, when, how, where and who questions are answered. Plans should be tested…

A checklist to test plans

	YES	NO
Have I called upon specialist advice?	☐	☐
Have all feasible courses of action been considered and weighed up in terms of resources needed/available and outcomes?	☐	☐
Has a programme been established which will achieve the objective?	☐	☐
Is there a provision for contingencies?	☐	☐
Were more creative solutions searched for as a basis for the plan?	☐	☐
Is the plan simple and as foolproof as possible, rather than complicated?	☐	☐
Does the plan include necessary preparation or training of the team and its members?	☐	☐

In ensuring that there is the appropriate level of participation in the planning process, the chart below may be useful:

The planning continuum

Use of authority by the leader

Area of freedom for team members

1. Leader makes a plan and announces it	2. Leader 'sells' own plan	3. Leader presents ideas and invites questions	4. Leader presents tentative plan subject to change	5. Leader presents problem, gets suggestions, makes plan	6. Leader defines limits; asks team to make plan

3 Briefing

Briefing or instructing a team is a basic leadership function conducted usually in a face-to-face way. Any briefing is an opportunity to:

- create the right **atmosphere**
- promote **teamwork**
- get to know, encourage and motivate each **individual**.

Before and after any briefing session, to ensure that the question of 'what is my role in all this?' (which will be on everyone's mind) is answered, you need to ask yourself these questions:

1 Does every individual know exactly what his/her job is?

2 Does each member of the team have clearly defined targets and performance standards agreed with me?

3 Does each person know at the end what is expected of him/her and how that contribution or that of his/her team fits in with the purposeful work of everyone else?

Communicating (speaking **and** listening) is crucial to get right in any briefing and it centres on the task, team and individual needs which should be addressed.

The effective speaking attributes of a successful briefing are to be:

- prepared
- clear
- simple
- vivid
- natural.

Assertiveness can be important. For example, to give the task direction and in explaining the role of the team/individual, especially in an initial briefing or where there is low morale.

In briefings, you could do worse than keep these points in mind:

The Adair short-course on leadership

1 The six most important words...   'I admit I made a mistake'.

2 The five most important words...   'I am proud of you'.

3 The four most important words ...   'What is your opinion?'

4 The three most important words...   'If you please'.

5 The two most important words...   'Thank you'.

6 The one most important word...   'We'.

7 The last, most **unimportant** word...   'I'.

4 Controlling

Excellent leaders get maximum results with the minimum of resources.

To control others, leaders need to exhibit self-control (but remembering that anger/sadness can be legitimate responses if the circumstances warrant it and are themselves mechanisms for control), to have good control systems (simple and effective to monitor financial and task performance) and to have control of what it is that others should and should not be doing in order to meet objectives. The success at directing, regulating, restraining or encouraging individual and team efforts on the task (and in meetings) are the criteria for testing a leader's effectiveness as a 'controller'.

A checklist for testing controlling skills

	YES	NO
Do I maintain a balance between controlling too tightly or giving too much freedom to the team?	☐	☐
Am I able to co-ordinate work-in-progress, bringing together all the parts in proper relation with each other?	☐	☐
In technical work, do I ensure that team and individual needs are met?	☐	☐
Do meetings I chair run over time(s) allotted to topics?	☐	☐
Do I have proper budgets and ways of monitoring actual performance?	☐	☐
Do customers rate my organisation's control systems for:		
– quality of product/service	☐	☐
– delivery	☐	☐
– costs	☐	☐
– safety?	☐	☐

5 Evaluating

Leaders need to be good at:

- assessing the consequences
- evaluating team performance
- appraising and training individuals
- judging people.

In *assessing the consequences*, leaders should be able to foresee the outcome of action (or inaction) in terms of the technical, the financial and the human and to ask probing questions of the team in order to establish the likely consequences.

In *evaluating team performance*, perhaps through a de-briefing session after a particular project, the performance of the team as a whole in relation to the task can be examined:

- Has it been a success, a partial success or a failure?
- Can lessons be learnt?
- Can action be taken to improve performance?
- What feedback can be given to **ensure** improvement?

The evaluation of the team is helpful in trying to build it into a high-performance one where the hallmarks are:

- clear realistic objectives
- shared sense of purpose
- best use of resources
- atmosphere of openness
- handles failure
- rides out the storms.

In *appraising and training individuals*, the following agenda can be used:

- Past performance
- Future work to be done: targets, priorities, standards and strategies
- Matching perceptions of what can be expected by each party of the other in order to achieve a good working relationship
- Improving skill, knowledge, behaviour.

Some tips in handling appraisals:

- Have all necessary data available
- Put the other person at ease
- Control pace and direction of the interview
- Listen... listen... listen
- Avoid destructive criticism (encourage self-criticism)
- Review performance systematically
- Discuss future action
- Discuss potential/aspirations
- Identify training/development required
- Avoid common pitfalls, such as:
  - dominating the conversation
  - making promises unlikely to be kept
  - expecting dramatic changes overnight
  - blaming those not present.

In *judging people*, leaders decide who should do what and this always affects outcomes and so is a crucial skill. Leaders should not have favourites because:

- it destroys team unity
- the favourite is a personification of your judgement about people – if others do not agree with your judgement, your credibility suffers
- favourites advance by recognising the social and esteem needs of their bosses and by pandering to them – the boss can have his/her judgement impaired by this.

Judgement is improved by analysing impressions formed, discussing them with others and by making decisions about people more slowly and after deliberation.

In evaluation, you need to ensure that:

- your decision-making judgement is good
- you appraise people regularly and well
- you are good at judging people
- you evaluate your own performance as much as those who work for you.

6 Motivating

There are six key principles for motivating others:

1 **Be motivated yourself**
2 **Select people who are highly motivated**
3 **Set realistic and challenging targets**
4 **Remember that progress motivates**
5 **Provide fair rewards**
6 **Give recognition.**

Individuals are motivated by their requirements to satisfy a (Maslow's) hierarchy of needs:

- Physiological – hunger, thirst, sleep
- Safety – security, protection from danger
- Social – belonging, acceptance, social life, friendship and love
- Esteem – self-respect, achievement, status, recognition
- Self-actualisation – growth, accomplishment, personal development.

Each individual will be at a different stage/level up this hierarchy of needs and will need to be motivated accordingly.

Other than in financial terms, individuals are usually motivated if they can see that they will be given:

- achievement
- recognition
- job interest
- responsibility
- advancement.

A good leader provides the right climate and the opportunities for these needs to be met on an individual basis and this is perhaps the most difficult of a leader's challenges.

Leaders must also inspire others. In 1987, James Kouzes and Barry Posner identified five characteristics of what they call exemplary leaders:

1 **Leaders challenge the process. Leaders search for opportunities. They experiment and take risks, constantly challenging other people to exceed their own limitations.**

2 **Leaders inspire a shared vision. Leaders envision an enabling future and enlist people to join in that new direction.**

3 **Leaders enable others to act. Leaders strengthen others and foster collaboration.**

4 **Leaders model the way. Leaders set the example for people by their own leadership behaviour and they plan small wins to get the process moving.**

5 **Leaders encourage the heart. Leaders regard and recognise individual contributions and they celebrate team successes.**

7 Organising

Good leaders are good at:

- organising themselves – their own work and particularly how they manage themselves, their time and how they delegate

- organising the team – to build and maintain it to ensure that there is good, effective team-work

- organising the organisation – the structure and the systems/processes in which, and by which, people operate.

Leaders change things and organise for the achievement of results – leading change requires considerable powers and skills of leadership. In all aspects, leaders must organise with a purpose clearly in mind at all times.

Leaders should consider their organising skills by reference to the Task, Team and Individual as follows:

- **Task:**   **is there a common purpose?**
    - is it broken down into aims and objectives?
    - how is it/are they communicated?
- **Team:**   **what are the teams/sub-teams?**
    - how do they contribute to the purpose?
    - do they relate together as a team?
- **Individual:**   **do they have freedom and discretion?**
    - are individual needs being met?

Further questions in surveying your organisation are:

- Do the Task/Team/Individual circles overlap sufficiently to provide and maintain high morale in the face of difficulties?
- How are tensions resolved and are there adequate systems /disciplinary procedures/dispute handling methods in place?

The size of working teams/groups should be examined to assess the importance of these factors:

- Task/technology – complexity narrows the span of control, ie. is the team too big to control/handle this aspect and does it mesh properly with other teams?
- Communications – especially with geographical/physical dispersement, are they good enough?
- Motivation and autonomy – is the training commensurate with any wishes to be self-sufficient?
- Competence of leaders – are large teams led by good enough leaders, what are the leader's other commitments and does he/she have good/specialist support?

A checklist to test the organising function ability

YOU   YES   NO

Can you organise your personal and business life in ways which would improve your effectiveness as a leader?

Do you delegate sufficiently?

Can you identify improvements in your time management?

TEAM

Is the size and make-up correct?

Should a sub-team be set up?

Are opportunities/procedures in place to ensure participation in decision-making?

Do you restructure and change individual's jobs as appropriate?

ORGANISATION

Do you have a clear idea of its purpose and how the parts should work together to achieve it?

Are effective systems in place for training/recruitment/ dismissal?

Do you carry out surveys into the size of teams, number of leadership levels, growth of unnecessary complexity, line and staff co-operation and properly working communications systems?

Are you good at administration, recognising the performance of administrators and ensuring that administrative systems facilitate excellent performance from teams/individuals?

8 Setting an example

'Leadership is example'. To be successful, a good leader must 'walk the talk'. Employees take a fraction of the time to know a leader as he/she takes to get to know them. The example you are giving is quite simply you. Whether this is a good or a bad example depends on the leader.

An example is set in verbal and non-verbal ways and all aspects of a leader's words and deeds must be considered in the light of this.

If example is contagious, it is worth ensuring that a good one is set to encourage the qualities sought in others.

Some key questions for good leadership are:

- **Task – do you lead from the front/by example?**
- **Team – do you develop your teams' standards through the power of example?**
- **Individual – do you view each individual as a leader in their own right?**

Bad example, particularly of hypocrisy, is noticed more than good, so care must be taken in all that a leader says and does.

A checklist to test if you set a good example

	YES	NO
Do you ask others to do what you would be unwilling to do yourself?	☐	☐
Do people comment on the good example you set in your work?	☐	☐
Does your (bad) example conflict with what all are trying to do?	☐	☐
Can you quote when you last deliberately set out to give a lead by example?	☐	☐
Can you think of ways you could lead by example?	☐	☐
Do you mention the importance of example to team leaders who report to you?	☐	☐

Developing leadership skills

Organisations (and if you are the leader of one that means you) should ensure that they have a policy of developing the leadership potential in all and particularly of newly appointed 'leaders'! Individuals should also ensure that they focus on developing their leadership skills by training, reading, analysing and following the example of good leaders and by assessing, monitoring and improving their own performance.

B. Teambuilding

This section looks at teambuilding from the leadership perspective and, as has been seen, teambuilding is part of the leadership 'holy' trinity of Task, Team and Individual.

One of the main results of good leadership is a good team:

Good leadership characteristics	Team outcomes
Enthusing	Team members are purposefully busy and have a basis to judge priorities
Lives values such as integrity	Gives a sense of excitement and achievement with people willing to take risks and higher work loads
Leads by example	Consistency in knowing leader's values
Generates good leaders from followers	Is trusted
Aware of own behaviour and environment	Aspire to leader's example
Intellect to meet job needs	Confidence in leadership
Aware of team and individual needs	The led start to lead (with leader less indispensable) being delegated to, coached and supported
Exhibits trust	Inspires confidence and performance
Represents the organisation to the team and vice versa	Confidence of contribution to aims and commitment to them

In 1985, ICI believed that the outcomes of effective leadership were that people will:

- have a clear sense of direction and work hard and effectively
- have confidence in their ability to achieve specific challenging objectives
- believe in and be identified with the organisation
- hold together when the going is rough
- have respect for and trust in managers
- adapt to the changing world.

In achieving the task, building the team and developing the individual, whilst leadership **style** may differ, effective leadership (in ICI's findings and its development courses) emphasized that the leader must do the following:

- feel personally responsible for his/her human, financial and material resources
- be active in setting direction and accepting the risks of leadership
- be able to articulate direction and objectives clearly and keep his/her people in the picture
- use the appropriate behaviour and methods to gain commitment for the achievement of specific objectives
- maintain high standards of personal performance and demand high standards of performance from others.

Leaders in teambuilding provide the functions of:

- planning
- initiating
- controlling

- supporting
- informing
- evaluating.

In relation to teambuilding and the leader's role in terms of the Task, Team and Individual (which need to be addressed if the Team is to perform at high levels of achievement) the following three sets of questions will help analyse and improve the way that teams operate.

1 Task

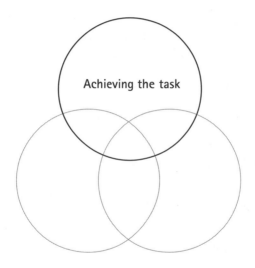

Purpose: Am I clear what the task is?

Responsibilities: Am I clear what mine are?

Objectives: Have I agreed these with my superior, the person accountable for the group?

Programme:	Have I worked one out to reach objectives?
Working conditions:	Are these right for the job?
Resources:	Are these adequate (authority, money, materials)?
Targets:	Has each member clearly defined and agreed them?
Authority:	Is the line of authority clear (Accountability chart)?
Training:	Are there any gaps in the specialist skills or abilities of individuals in the group required for the task?
Priorities:	Have I planned the time?
Progress:	Do I check this regularly and evaluate?
Supervision:	In case of my absence who covers for me?
Example:	Do I set standards by my behaviour?

2 Team

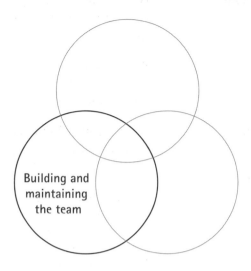

Building and
maintaining
the team

Objectives:	Does the team clearly understand and accept them?
Standards:	Do they know what standards of performance are expected?
Safety standards:	Do they know consequences of infringement?
Size of team:	Is the size correct?
Team members:	Are the right people working together? Is there a need for sub-groups to be constituted?
Team spirit:	Do I look for opportunities for building teamwork into jobs? Do methods of pay and bonus help to develop team spirit?

Discipline:	Are the rules seen to be reasonable? Am I fair and impartial in enforcing them?
Grievances:	Are grievances dealt with promptly? Do I take action on matters likely to disrupt the group?
Consultation:	Is this genuine? Do I encourage and welcome ideas and suggestions?
Briefing:	Is this regular? Does it cover current plans, progress and future developments?
Represent:	Am I prepared to represent the feelings of the group when required?
Support:	Do I visit people at their work when the team is apart? Do I then represent to the individual the whole team in my manner and encouragement?

3 Individual

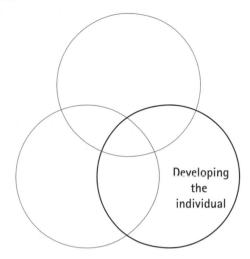

Developing
the
individual

Targets: Have they been agreed and quantified?

Induction: Does s/he really know the other team members and the organisation?

Achievement: Does s/he know how his/her work contributes to the overall result?

Responsibilities: Has s/he got a clear and accurate job description? Can I delegate more to him/her?

Authority: Does s/he have sufficient authority for his/her task?

Training: Has adequate provision been made for training or retraining both technical and as a team member?

Recognition:	Do I emphasise people's successes? In failure is criticism constructive?
Growth:	Does s/he see the chance of development? Does s/he see some pattern of career?
Performance:	Is this regularly reviewed?
Reward:	Are work, capacity and pay in balance?
The task:	Is s/he in the right job? Has s/he the necessary resources?
The person	Do I know this person well? What makes him/her different from others?
Time/attention	Do I spend enough with individuals listening, developing, counselling?
Grievances:	Are these dealt with promptly?
Security:	Does s/he know about pensions, redundancy and so on?
Appraisal:	Is the overall performance of each individual regularly reviewed in face-to-face discussion?

The good leader in teambuilding must act as:

- encourager
- harmoniser
- compromiser
- expediter/gatekeeper

- standard setter
- group observer/commentator
- follower.

As leader, there must be a clear understanding of:

i) Team properties

- common background/history (or lack of it)
- participation patterns
- communication
- cohesiveness
- atmosphere
- standards
- structure
- organisation
- changes over time (forming, storming, norming and performing) both progressive and regressive
- how to change the team properties in evidence.

ii) Team roles being defined, but with room left for individual personality

iii) Team member functions

- distinction between content (the what) and process (the how) of group functioning
- difference between behaviour related to the task and behaviour related to maintenance of the team and that behaviour which expresses individual idiosyncrasies
- team leader functions (as above).

iv) The individual

- balancing interests and self-expression of individuals and the team
- the value of the task draws individuals/team together
- having sound values motivates individuals in teams.

v) The overlapping needs of Task, Team and Individual need to be addressed

vi) Team processes

- to see what is really going on
- improved decision-making rests on seeing beneath the surface the pressures that influence the team
- calmness creates interdependence within the team and with the leader
- avoid team flight into abstractions
- aim for consensus (only where possible)
- assess team view of authority to see how processes/decisions are being affected by it.

vii) Teams within teams

- watch out for hostility, communication failure and mistrust as signs of team fragmentation
- develop teamwork between teams as well as within them
- winning can be as destructive to teams as losing, if not worse, unless both outcomes are handled well
- be aware of teams within teams and act accordingly to regain cohesiveness or sub-divide.

Summary and six-month follow up test

Summary

Leadership centres on:

- the leader – qualities of personality and character
- the situation – partly constant, partly varying
- the team – the followers: their needs and values
- the overlapping needs of the Task, Team and Individual
- leadership functions can be summarised as:
    - defining the task
    - planning
    - briefing
    - controlling
    - evaluating
    - motivating
    - organising
    - providing an example

Teambuilding centres on:

- achieving the task
- building and maintaining the team
- developing the individual.

Six-month follow up test

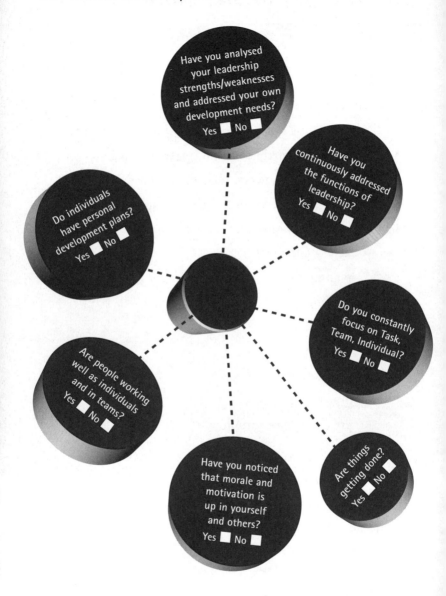

Have you analysed your leadership strengths/weaknesses and addressed your own development needs?
Yes ☐ No ☐

Have you continuously addressed the functions of leadership?
Yes ☐ No ☐

Do individuals have personal development plans?
Yes ☐ No ☐

Do you constantly focus on Task, Team, Individual?
Yes ☐ No ☐

Are people working well as individuals and in teams?
Yes ☐ No ☐

Have you noticed that morale and motivation is up in yourself and others?
Yes ☐ No ☐

Are things getting done?
Yes ☐ No ☐

07

MOTIVATION AND PEOPLE MANAGEMENT

07

Introduction

Getting the best from people, achieving results through individuals and teams, maintaining consistent high performance, inspiring oneself and others into action – all depend on the skills of motivation. Self-motivation can be as difficult as motivating others and you cannot have one without the other.

Adair's eight rules in motivating people

1. Be motivated yourself
2. Select people who are highly motivated
3. Treat each person as an individual
4. Set realistic and challenging targets
5. Remember that progress motivates
6. Create a motivating environment
7. Provide fair rewards
8. Give recognition

Understanding what moves an individual to action is crucial in a manager being able to engage the will to act. Motives (which operate the will which leads to action) are inner needs or desires and these can be conscious, semi-conscious or unconscious. Motives can be mixed, with several clustered around a primary motive.

The 50:50 rule

Just as the Pareto principle (or 80:20 rule) is the ratio of 'the vital few and trivial many', the Adair 50:50 rule has it that:

50% of motivation comes from within a person; and

50% from his or her environment, especially from the leadership encountered therein.

Unfortunately human behaviour and what decides/triggers it is more complicated than the carrot and stick 'theory' which deals only with external stimuli. The 'carrot' of reward/incentive and the 'stick' of fear of consequences reveal only two 'motives' which govern action. There are many more!

The expectancy theory – formulated by Edward C Tolman in the 1930s (whereby behaviour rests on the instinctive tendency for individuals to balance the value of expected benefits against the expenditure of energy) falls into the same 'stimulous-response' approach to motivation. It does demonstrate, however, that an individual's strength of motivation can be affected by the expectations of outcomes from certain actions *and* further strengthened by the individuals preferred outcome, as demonstrated by Victor H. Vroom in the 1960s.

It pays, therefore, in external stimuli to bear in mind that:

1 the routes to desired outcomes for individuals and teams are clear; and

2 individuals perceive the rewards or punishments in different ways according to their own values.

This confirms the need to treat people as individuals but as the 50:50 rule also indicates, other motivational factors should always be set in the context of the individual's managed environment. Other theories of motivation which suggest that 90% of motivation is within an individual should be tempered by the 50:50 rule.

Needs and motivation

Maslow's hierarchy of needs

A sketch map of individual needs – which is useful for managers when considering individuals – can be drawn from Maslow's hierarchy of needs (1954), but it must be borne in mind that his theory does not fully appreciate individual differences or that each person has a unique set of needs and values.

Maslow identified five motivating factors in his hierarchy of needs and indicated that as each need is satisfied, others then emerge. He identified:

1   **physiological needs (including hunger, thirst, sleep)**

2   **safety needs (security and protection from danger)**

3   **social needs (belonging, acceptance, social life, friendship and love)**

4   **self-esteem (self-respect, achievement, status, recognition)**

5   **self-actualisation (growth, accomplishment, personal development).**

However, points to bear in mind are that:

- individuals do not necessarily move up the hierarchy on the principle that a 'satisfied need ceases to motivate' although that can be the case

- different levels of needs can kick in at random points on the scale toward full satisfaction of needs

- culture and age and other factors can affect the importance of the different needs to different people and at different stages in their lives

- the satisfying of some needs can be sacrificed in order to try and satisfy higher level needs.

McGregor's Theory X and Theory Y

In 1960 in his book 'The Human Side of Enterprise', McGregor demonstrated that the way in which managers manage depends on the assumptions made about human behaviour. He grouped these assumptions into Theory X and Theory Y.

Theory X – the traditional view of direction and control

i) The average human being has an inherent dislike of work and will avoid it if possible.

ii) Because of this dislike of work, most people must be coerced, controlled, directed, threatened with punishment to get them to give adequate effort toward the achievement of organisational objectives; and

iii) The average human being prefers to be directed, wishes to avoid responsibility, has relatively little ambition and wants security above all.

Theory Y – the integration of individual and organisational goals

(i) The expenditure of physical and mental effort in work is as natural as play or rest;

(ii) External control and the threat of punishment are not the only means for bringing about effort toward organisational

> objectives. People will exercise self-direction and self-control in the service of objectives to which they are committed;
>
> (iii) Commitment to objectives is a function of the rewards associated with their achievement;
>
> (iv) The average human being learns, under proper conditions, not only to accept, but to seek responsibility;
>
> (v) The capacity to exercise a relatively high degree of imagination, ingenuity and creativity in the solution of organisational problems is widely, not narrowly, distributed in the population.
>
> (vi) Under the conditions of modern industrial life, the intellectual potentialities of the average human being are only partially utilised.

McGregor drew on Maslow for much of Theory Y and put forward the cluster of features as an unproven hypothesis and further research was needed (Herzberg) to seek to prove it correct.

In terms of management in practice Theory Y does reveal that in any individual within an organisation there are untapped resources of goodwill, energy, creativity and intelligence.

Herzberg's Motivation – hygiene theory

In Herzberg's research (published in his 1959 book 'The Motivation to Work'), fourteen factors were identified to be the sources of good or bad feelings:

1 Recognition

2 Achievement

3 Possibility of growth

4 Advancement

5 Salary

6 Interpersonal relations

7 Supervision – technical

8 Responsibility

9 Company policy and administration

10 Working conditions

11 Work itself

12 Factors in personal life

13 Status

14 Job security

The eight **'hygiene'** factors, according to Herzberg, which can create job dissatisfaction are:

1 **Company policy and administration**
   - availability of clearly defined policies, especially those relating to people
   - adequacy of organisation and management

2 **Supervision – technical**
   - accessibility, competence and fairness of your superior

3 **Interpersonal relations**
   - relations with supervisors, subordinates and colleagues
   - quality of social life at work

4 **Salary**
   - total compensation package, such as wages, salary, pension, company car and other financially related benefits

5   **Status**
    - a person's position or rank in relation to others, symbolised by title, size of office or other tangible elements

6   **Job security**
    - freedom from insecurity, such as loss of position or loss of employment altogether

7   **Personal life**
    - the effect of a person's work on family life, eg stress, unsocial hours or moving house

8   **Working conditions**
    - the physical conditions in which you work
    - the amount of work
    - facilities available
    - environmental aspects, eg ventilation, light, space, tools, noise

The six **motivating** factors that lead to job satisfaction were identified by Herzberg as being:

1   **Achievement**
    - specific successes, such as the successful completion of a job, solutions to problems, vindication and seeing the results of your work

2   **Recognition**
    - any act of recognition, whether notice or praise (separating recognition and reward from recognition with no reward)

3   **Possibility of growth**
    - changes in job where professional growth potential is increased

4 **Advancement**
   - changes which enhance position or status at work

5 **Responsibility**
   - being given real responsibility, matched with necessary authority to discharge it

6 **The work itself**
   - the actual doing of the job or phases of it.

The **hygiene** factors are those where people seek to avoid particular situations, whereas the **motivating** factors are matched with people's needs to achieve self-actualisation or self-realisation.

Satisfaction of the Herzberg motivators and avoidance of problems with the hygiene factors can help you as a manager to assess roles and jobs within your organisations to check what job-enrichment or empowerment you ought to contemplate to improve performance and give individuals greater job satisfaction.

Managers/leaders and motivation

Managers and leaders should take a realistic and visionary view of people who work for them and with them. Individuals can be managed better if it is recognised that they are:

1 individuals, but become fully developed and truly themselves in relation to other people and meaningful work

2 creative and imaginative, but only in concert with others through working on their own or in teams

3 driven by achievement (as individuals) but know that they achieve more as part of a team

4   self-motivated and self-directed but need management/ leadership (if only to co-ordinate activities)

5   intelligent enough to know the difference between rewards such as money and those less tangible rewards that meet value needs

6   interested in leaving work/the world a better place and know that that yields a bonus.

As has been described in the relevant section earlier in the book, in leadership, the achievement of the task, the building and maintaining of the team and the development of the individual can only result from motivating people by providing the leadership functions of:

- planning
- initiating
- controlling
- supporting
- informing
- evaluating

and by being able to inspire others.

Managers should check that individuals have:

1   a sense of achievement in their job and feel that they are making a worthwhile contribution to the objective of the team

2   jobs which are challenging and demanding with responsibilities to match capabilities

3   adequate recognition for achievements

4   control over delegated duties

5   a feeling that they are developing along with growing experience and ability.

Manager's motivating checklist

DO YOU

	YES	NO
Agree with subordinates their main targets and responsibilities together with standards of performance, so that you can both recognise achievement?	☐	☐
Recognise the contribution of each member of the team and encourage team members to do the same?	☐	☐
Acknowledge success and build on it?	☐	☐
Analyse set-backs, identifying what went well and giving constructive guidance to improve future performance?	☐	☐
Delegate as much as possible giving more discretion over decisions and accountability to a sub-group or individual?	☐	☐
Show those that work with you that you trust them or do you surround them with unnecessary controls?	☐	☐
Provide adequate opportunities for training and re-training if necessary?	☐	☐
Encourage each individual to develop his/her capacities to the full?	☐	☐
Review the overall performance of each individual regularly face-to-face?	☐	☐
Match financial reward to contribution?	☐	☐
Make time to talk and listen so that you understand the unique and changing profile of needs and wants in each person and work with what you find?	☐	☐
Encourage able people with the prospect of promotion within the organisation?	☐	☐

Getting the best from people

Be motivated yourself

Enthusiasm and motivation inspires others and the badges of good example setting are that you should be:

- public – make sure you act in the open
- spontaneous – do not appear calculated
- expressive – do things because they are natural to you not for effect
- self-effacing – setting a good example is not glory-seeking.

Motivation is contagious so you should be infectious! If you are not motivated yourself, you cannot motivate others and you should examine the reasons why you lack motivation. Symptoms include having little or no interest in the job, wanting to arrive late and leave early, wanting to leave the job, feeling active dislike for it and feeling out of place in it.

You can strengthen your motivation by reminding yourself:

1 to feel and act enthusiastically and in a committed way in your work

2 to take responsibility when things go wrong rather than blaming others

3 to identify ways you can lead by example

4 to act on the 50:50 principle

5 to motivate by word and example rather than manipulation

6 to set an example naturally rather than in a calculated way

7 not to give up easily

8 to ensure you are in the right job for your own abilities, interests and temperament

9 to be able to cite experiences where what you have said or done has had an inspirational effect on individuals, the team or the organisation

10 that the three badges of leadership are enthusiasm, commitment and perseverance.

Select people who are highly motivated

The seven key indicators of high motivation are:

1 **energy** – not necessarily extrovert but alertness and quiet resolve

2 **commitment** – to the common purpose

3 **staying power** – in the face of problems/difficulties/set-backs

4 **skill** – possession of skills indicates aims and ambitions

5 **single-mindedness** – energy applied in a single direction

6 **enjoyment** – goes hand in hand with motivation

7 **responsibility** – willingness to seek and accept it.

Choosing people well means looking at motivation, ability and personality and you should, when interviewing, look for real evidence behind the interviewee's facade.

Looking for the Michelangelo motive (where the quality of the work itself is a key motivator) can yield good results in selecting highly motivated individuals. You should look for:

- a sense of pride in the individual's own work

- an attention to detail

- a willingness to 'walk the extra mile' to get things right

- a total lack of the 'its good enough, let it go' mentality

- an inner direction or responsibility for the work (without the need for supervision)
- an ability to assess and evaluate own work, independently from the opinions of others.

It should be stressed that perfectionism is not what is called for – the best can be the enemy of the good.

Managers should check whether individuals are in the right job with the right skills and abilities, otherwise motivation techniques will fail. The aim is to select people who are motivated for the most appropriate job.

Treat each person as an individual

Find out what motivates an individual, do not rely on generalised theories or assumptions. Enter into a dialogue with each team member – help them to clarify what it is that motivates them – and use what you find to mutual benefit.

In each person you should engender a sense of:

- trust
- autonomy
- initiative
- industry
- integrity
- security.

Take time with each individual to:

- encourage
- hearten
- inspire

- support
- embolden
- stimulate.

Ask yourself these questions to ensure that you treat each person as an individual:

1   Do you know the names of the people on your team and their teams if they are leaders?

2   Can you identify ways in which those who report to you differ from each other?

3   Do you accept that an individual's motivation changes from time to time?

4   Do you spend time with people to know them, work with them, coach them?

5   Does your organisation see you as an individual?

Set realistic and challenging targets

This can only be done in the context of understanding the organisation's aims or purpose. It is only then that targets and objectives can be identified and tasks defined.

Jacob's ladder model

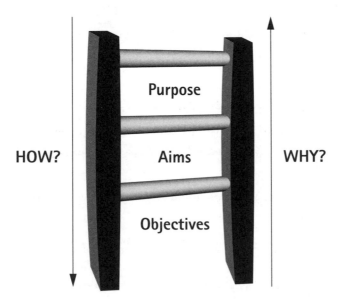

Moving **down** the ladder, ask **how** are we going to achieve the task? The answer is by breaking down the purpose into the main aims and the main aims into short and long-term objectives or goals.

Moving **up** the ladder, ask **why** are we doing this? The answer is to achieve this objective in order to achieve this aim and to satisfy this purpose.

Targets set (for short and longer-term objectives) should be:

- specific
- clear
- time-bounded.

An objective or target must be realistic (feasible) and challenging. If you aim for the best you can often get it.

Targets must be agreed and monitored with further action agreed to maintain motivation toward shared objectives.

Remember that progress motivates

There is a 'lust to finish' (John Wesley) and the key principle is that progress motivates – moving forward leads them to raise their game.

Feedback on progress (or even the relative lack of it) helps motivation either to spur people on or to concentrate the mind on what yet needs to be done.

Feedback is not given at all or sometimes not often enough, usually for these reasons:

- 'People don't need to be told how they are doing, they already know'
- 'People take it easy if you say things are going well'
- 'They are unhappy and cause trouble if you say things are not going well'
- 'We lack the skills or the time to do it'.

Feedback which is affirmative (praise) must be:

- accurate
- sincere
- generous
- spontaneous
- fair.

It must not be:

- patronising
- superior/condescending
- grudging
- calculated for effect.

Maintaining motivation depends on informing and inspiring and the rule is always to give information first, before you attempt to encourage.

Maintaining high morale is key to high motivation and morale covers the individuals and the team. Where an individual has low morale, the issues have to be addressed on an individual basis, but where group or team morale is low, the answer lies in deciding whether there is a lack of confidence:

- of ultimate success
- in the present plan(s)
- in the leadership/management
- in the minds of team members.

It can be necessary to remotivate the team by rebuilding self-confidence and by readdressing:

- aims – and clarifying objectives
- plans, resources needed
- leadership
- overlooked factors
- re-establishing the worth or value of the task(s)
- involvement of individuals in key decisions.

Create a motivating environment

Guidelines here are:

1 Beware of creating a restrictive organisation with an over-emphasis on controls

2 Avoid public criticisms of individuals

3 Ensure Herzberg's hygiene factors are catered for – the physical and psychological well-being of people should have high priority

4 Control systems should only be introduced where necessary

5 Give people an input into decisions which affect their working lives (especially in respect of substantial change)

6 Keep units and sub-units as small as possible (for large units tend to be bureaucratic and demotivational if they lack inspired leaders)

7 Pay attention to job design – avoid repetitive work, introduce variety

8 Give people autonomy and a job with a 'product' that an individual can recognise as his/her own

9 Ensure an individual understands the significance of their job in relation to the whole, which will also encourage new ideas and innovation.

Provide fair rewards

Although it is difficult to ensure that the financial reward an individual receives is fair (commensurate with contribution), effort must be applied in trying to get it right. There are other motivating 'returns' that individuals look for from jobs (as in Maslow's hierarchy of needs), but money is the one which has the main strategic importance for most people.

Most individuals like the combination of a fixed salary with a variable element related to performance or profits.

Also of tactical importance are incentives to improve performance in key areas, eg sales, customer service and credit control.

Incentives can be in the form of cash, vouchers, merchandise or travel, but care must be taken to administer any incentive schemes fairly and without risking demotivating any 'losers'.

In providing fair rewards, an organisation should ask itself:

1 Do we have a scheme whereby financial reward is made up of a fixed and variable element?

2 Do we link performance and pay?

3 Have we addressed the problems of whether to pay performance-related elements to the team or the individual?

4 Do we actively consider changing our information systems to improve methods of rewarding performance?

5 Do we have schemes other than for sales people?

6 Does our organisation reward the behaviours/performance that publicly it values?

7 Do senior managers have pay rises/bonuses when they expect others to do without them?

It is always worth remembering Herzberg's insight that salary has more power to make people dissatisfied or unhappy than it has the power to motivate them.

Give recognition

Financial reward is seen by the recipient as a tangible form of recognition. There are other ways whereby appreciation is expressed for what has been contributed.

If recognition is not given, an individual can feel unnoticed, unvalued and unrewarded. This leads to a drop in motivation and energy levels.

Recognition should be formal or informal, for the individual and/or the team, as appropriate.

In giving recognition, you should try to ensure that you:

1 treat everyone in a fair and equal way
2 reward real achievements or contributions
3 reflect the core values of the organisation
4 use it to guide and encourage all concerned
5 give it in public if possible
6 give it formally and informally
7 give it genuinely and sincerely.

Other than financial payments, any words of recognition could be reinforced by giving:

- time off (with pay)
- tickets for an event or dinner out
- small gift
- special project of importance
- a change in job title.

It is a good idea to back up words of praise or recognition with some tangible gift.

Find out what is going on, share praise received with subordinates, and say thank-you more often, because people really value positive recognition and are motivated by it.

Know people's names – that is the basic form of recognition!

Summary and six-month follow-up test

To draw the best out of people you need to:

- be motivated yourself
- select people who are already motivated
- set challenging but realistic targets
- remember that progress motivates
- treat each person as an individual
- provide fair rewards
- give recognition.

Six-month follow-up test

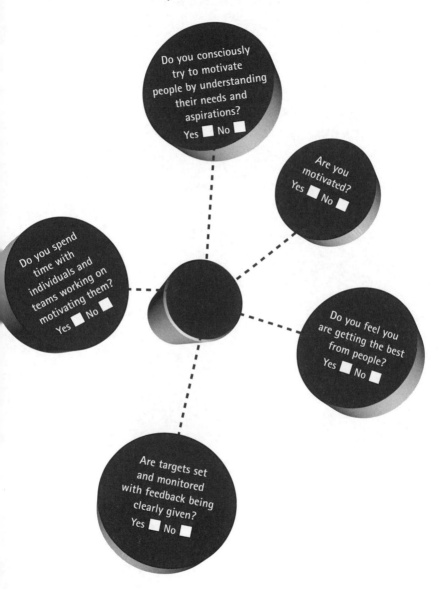

Do you consciously try to motivate people by understanding their needs and aspirations?
Yes ☐ No ☐

Are you motivated?
Yes ☐ No ☐

Do you spend time with individuals and teams working on motivating them?
Yes ☐ No ☐

Do you feel you are getting the best from people?
Yes ☐ No ☐

Are targets set and monitored with feedback being clearly given?
Yes ☐ No ☐

08

COMMUNICATION AND PRESENTATION

08

Introduction

It is self evident that written and spoken communication skills are of crucial importance in business (and personal) life. Managers and leaders in particular must be effective communicators, good at getting their message across to, and at drawing the best out of, people. Communication skills in all forms, including non-verbal communication, need to be worked at and improved to ensure you understand people and they understand you.

Issues in communication

- You must be in social contact with the other person or people
- You must want to communicate
- It is better to risk familiarity than be condemned to remoteness
- The best way to empower others is to impart information (along with the delegated authority to make decisions and act on the information given)
- Get out of your office – meet, listen, provide information and give people the context in which they operate – to communicate and encourage
- Good communication is the core of customer care
- Remember customers (and suppliers) communicate with others about you
- To communicate with your customers you must handle complaints (as an organisation) as personally as possible – by a meeting or phone call in preference to letter or fax; you must listen to what customers suggest and communicate product/service changes/developments with them in advance

- Presentation skills are important in communicating with colleagues as well as customers/clients

- Meetings, internal and external are key indicators of a person's communication (including listening) skills

- Communication is a business requirement: establish proper systems and ensure all use them

- Remember the equation: size + geographical distance = communication problems

- Communicate with poor performers to improve their contribution and in appraisals be truthful, helpful and tactful

- Help others to improve their communication skills

- Assess your own communication skills and strive to improve them bit by bit. (Also, assess the communication skills of colleagues and identify areas for improvement.)

Personal communication skills checklist

	YES	NO

Do you understand the importance of communication in your personal and business life? ☐ ☐

Are you a good communicator? (Check with your partner at home, with friends and with colleagues.) ☐ ☐

Can you write down your strengths and weaknesses as a communicator? And have you listed them? ☐ ☐

Have you identified a need to improve your communication skills in any or all of these areas and will you now set about doing so (reading further books and/or attending training seminars as needs be):

- listening ☐ ☐
- reading ☐ ☐
- writing ☐ ☐
- one-to-one interviews ☐ ☐
- speaking and presentation ☐ ☐
- managing meetings ☐ ☐
- within your organisation? ☐ ☐

Are you motivated strongly to become an excellent communicator? ☐ ☐

Listening

Listening has been called the forgotten skill in communication. It is more than just hearing, it is the giving of thoughtful attention to another person whilst they are speaking.

The 'disease of not listening' – the 'I hear what you say' response – exhibits the following symptoms:

- Selective listening is habit forming: not wanting to know things and turning a deaf ear to certain types of information does two things:

    i) you do not listen to important items

    ii) people censor what they tell you

    and both can be damaging in business and in your private life

- The constant interrupter is not listening (but planning his/her own next interruption)

- The 'day-dreamer' is not a listener

- The poor listener is easily distracted by external factors, eg noise, heat/cold

- The lazy listener makes no effort with difficult information

- The poor listener over-reacts to a speaker's delivery and/or quality of visual aids rather than concentrating on what is being said.

The tell-tale signs of a good listener:

- Paying close attention to others when they are talking
- Taking an interest in someone you meet for the first time, trying to find an area of mutual interest
- Believing everyone has something of value to teach or impart to you
- Setting aside a person's personality/voice in order to concentrate on what they know
- Being curious in people, ideas and things
- Encouraging a speaker (with nods or eye contact)
- Taking notes
- Knowing one's own prejudices and working at controlling them to ensure listening continues
- Being patient with poor communicators
- Not being told you don't listen
- Having an open mind in respect of other peoples' points of view

Listening skills centre on the five following attributes:

1 **Being willing to listen**
2 **Clearly hearing the message**
3 **Interpreting the meaning (the speaker's meaning, not only your interpretation)**
4 **Evaluating carefully (suspending judgement at first but then assessing value and usefulness)**
5 **Responding appropriately – remembering communication is a two-way street.**

In active listening you must be prepared to:

- ask questions
- weigh the evidence
- watch your assumptions
- listen between the lines (at what is not said *and* for non-verbal elements such as facial expressions, tone, posture, physical gestures etc.)

Reading skills

Good reading is listening in action – giving time and thought and remaining alert to the possibilities suggested. A good reader will try to work past:

- poor structure and layout
- boring style
- off-putting tone
- too much or too little information
- difficult to follow content
- inordinate length
- lack of illustrations/diagrams.

You should examine what materials you **must** read, **should** read or **might** read in the light of your job/role/future ambitions and then decide accordingly how and when to handle a particular item.

Speed reading is useful but only if it is accompanied by speed under-standing and reading too fast (or too slowly) can impair understanding.

Read selectively (according to the must, should or might categorisation) from each item that confronts you. In this, scanning can help decide what attention to give particular items, so you should look at overall content (headings and sub-headings), sample the style and content of a few paragraphs, scan (if still interested) selected parts and then read that which you decide you are interested in. In reading carefully, you should be aware of the need to:

- be clear about your purpose of reading any piece of writing
- have questions in mind
- keep the questions firmly in mind and seek answers to them
- read for main ideas
- test the evidence, explanations and conclusions critically
- make notes as you progress
- test the writer's experience against your own
- consider whether or not to re-read
- discuss the material with others if appropriate
- reflect on what has been read.

Writing skills

Communicating in writing is an essential part of your job. There are three key elements in communicating on paper:

- Structure and lay-out
- Content
- Style and tone.

Writing should be thought of as talking to a person on paper and the six principles of good spoken communication apply – and they are:

1 **Clarity**

2 **Planning and preparation**

3 **Simplicity**

4 **Vividness**

5 **Naturalness**

6 **Conciseness.**

In letters, reports and memos the quality improves if the appropriate amount of planning is given to the points you wish to make and their order of importance. Further drafts can improve on the initial effort.

In writing a business letter you should always test the draft to ensure that:

1 the message is clear

2 points are made in the best order

3 it has the right style and tone

4 the most appropriate words and phrases are being used

5 the grammar/spelling is correct

6 the layout is attractive.

In writing reports which work the following points should be borne in mind:

- If the report is to stand alone and not to support a briefing or presentation it will need to be more than an aide-memoire

- A report should:

    – have an introduction with background and objectives

- a title which indicates its purpose
- be structured like a book with chapters, headings and sub-headings all clearly numbered and signposted
- ensure the main body of evidence is succinct and arranged in an easy to follow order
- end off with conclusions and recommendations
- indicate assumptions made
- put complicated data into an appendix
- use illustrations/diagrams to clarify points made

- Easy reading makes hard writing

- Churchill's guidelines for report writing centred on:
  - setting out main points in a series of short, crisp paragraphs
  - complicated factors or statistics should form an appendix
  - considering submitting headings only, to be expanded orally
  - avoiding woolly phrases, opting for conversational phrases
  - setting out points concisely aids clearer thinking

- Reports can be tested for their effectiveness as follows:
  - is the structure and layout clear and easy to follow?
  - is the content complete and does it:
    - state the purpose?
    - say when, by whom, for whom and with what scope it was prepared?
    - identify and address the problem clearly?

- ensure detail does not cloud the main issue?
- give sources for facts?
- use consistent symbols and abbreviations?
- use accurate figures?
- make clear statements?
- have conclusions which flow logically from facts and their interpretation?
- ensure other possible solutions are only abandoned with stated reasons?
  - in general
    - is the report objective?
    - are criticisms of its recommendations pre-empted?
    - is it efficient and business-like?
  - does it offend anyone?
  - can it be understood by a non-technical person?
  - is it positive and constructive?
  - does it point up the decision to be made and by whom?

The style and tone of written communications is important to ensure the message is put over, and received, clearly. Some rules are:

- keep it simple
- strive for clarity above all things (even above brevity)
- be natural
- be concise
- let the tone reflect your true feelings but beware of being terse, curt, sarcastic, peevish, angry, suspicious, insulting, accusing, patronising or presumptious
- be courteous (cordial and tactful).

Speaking and presentation skills

Effective speaking

There are certain principles to be followed to increase the power of communicating or expressing thoughts in spoken words.

Adair's six principles of effective speaking

1. Be clear
2. Be prepared
3. Be simple
4. Be vivid
5. Be natural
6. Be concise

Preparation is helped by asking the Who? What? How? When? Where? Why? of the speaking occasion to focus on the audience, the place, the time, the reasons giving rise to the occasion, the information that needs to be covered and how best to put it across.

Presentation skills

There are six clusters which form the main elements of good, effective presentation skills.

1 Profile the occasion, audience and location

You should ask yourself these questions:

- The occasion
  - what kind is it?
  - what are the aims of it?
  - what time is allowed?
  - what else is happening?

- The audience
  - do they know anything about you?
  - do you know its size?
  - what do they expect?
  - why are they there?
  - what is their knowledge level?
  - do you know any personally/professionally?
  - do you expect friendliness, indifference or hostility?
  - will they be able to use what they hear?

- The location
  - do you know the room size, seating arrangements, lay-out/set-up and acoustics?
  - do you know the technical arrangements for use of microphones, audio-visuals, lighting and whether assistance is available (and have you notified in advance your requirements)?
  - do you know who will control room temperature, lighting and moving people in and out?
  - have you seen it/should you see it?

2   Plan and write the presentation

Elements to address are:

- Deciding your objective which needs to be
  - clear
  - specific
  - measurable
  - achievable in the time available
  - realistic
  - challenging
  - worthwhile
  - participative

- Making a plan with a framework which has
  - a beginning (including introductory remarks, statement of objectives and relevance and an outline of the presentation(s))
  - a middle (divided into up to six sections maximum, ensuring main points are illustrated and supported by examples or evidence, use summaries and consider time allocation carefully – and test it)
  - an end (summarise, linking conclusions with objectives and end on a high note)

3   Use visual aids

As up to 50 per cent of information is taken in through the eyes, careful consideration should be given to the **clear**, **simple** and **vivid** use of audio-visuals.

Useful tips are:

- Powerpoint slides help make a point and keep eye contact with an audience (look at the people not the slides)

- Only present essential information in this way (keep content to about 25 words or equivalent if in figures)

- Have them prepared with appropriate professionalism
- Know the order
- Use pictures and colour if possible
- Do not leave a visual aid on for too long

Some difficulties with the different types of audio-visual equipment are:

- Powerpoint slides: make sure you are fully aware of how to use the equipment before you begin

- 35mm slide projection: professional in appearance, good for large audience and easy to use with a remote control can be offset by the need for dim lights (making note-taking difficult) and lack of flexibility in changing order of viewing

- Flipcharts: are easy to use and informal but difficult to use successfully with large groups and generally do not look professional and take up time to use

- Computers/tape decks/videos: can provide variety but difficult to set-up and synchronise, especially without technical support

4   Prepare your talk

In preparing your talk you need to decide whether you are to present with a full script, notes or from memory. This depends on the occasion and purpose of the presentation but whichever method is chosen, it is always acceptable to refer to your fuller notes if needs be during a presentation. Notes on cards or on slides/ flipcharts can be used as memory joggers if you present without notes. If you are required to read a paper, at least be able to look up occasionally. Remember that failing to prepare is preparing to fail.

5 Rehearse with others

Rehearsal is important, but not so much that spontaneity is killed and naturalness suffers, to ensure the presentation (and any audio-visual aid) is actually going to work in practice.

You should always visit the location if at all possible and check that everything works – knowing the location is as important as rehearsing the presentation, indeed it is an essential part of the rehearsal.

6 Delivery on the day

Overall you should ensure that your presentation's:

- **beginning** – introduces yourself properly, captures the audience and gives the background, objectives and outline of your talk.

- **middle** – is kept moving along (indicating whether questions are to be asked as-you-go or at the end) with eye contact over the whole audience, at a reasonable pace, with a varying voice and obvious enjoyment on your part.

- **end** – is signalled clearly and then goes off with a memo-rised flourish.

- **questions** – are audible to all (or repeated if not), answered with conciseness, stimulated by yourself asking some questions, dealt with courteously and with the lights on.

- **conclusion** – is a strong summary of talk and questions/discussions and closes with words of thanks.

If you find you are nervous (and this is normal) experiencing fear and its physical manifestations, remember to:

1 Breathe deeply

2 Manage your hands

3 Look at your audience

4 Move well

5 Talk slowly

6 Compose and relax yourself

7 Remember that the audience is invariably on your side

8 Project forward to the end of the presentation and picture the audience applauding at the end.

One-to-one interviews

Such meetings have the common characteristics that they are (usually) pre-arranged, require preparation and have a definite purpose.

Unless it happens to be a dismissal, one-to-one interviews require that:

- both parties know the purpose of the meeting (notified in advance)

- information to be exchanged should be considered in advance and answers at the meeting should be honest

- as interviewer you should keep control: stick to the point at the issue and the time allocated and give the other party adequate time to talk (prompting by questions if necessary).

The structure of the interview should be as follows:

- the opening – setting the scene, the purpose and a relaxed atmosphere
- the middle – stay with the purpose, listen, cover the agenda
- the close – summary, agree action, end naturally not abruptly on a positive note.

Sometimes it is useful to ask the right questions to obtain the required information/exchange. Questions to use are the open-ended, prompting, probing, or what-if questions, whilst the ones to avoid (unless being used for specific reasons) are the yes/no, closed, leading or loaded questions.

In performance appraisal interviews the aim should be to give constructive criticism in the following way:

1 **In private**
2 **Without preamble**
3 **Simply and accurately**
4 **Only of actions that can be changed**
5 **Without comparison with others**
6 **With no reference to other people's motives**
7 **Without apology if given in good faith.**

In receiving constructive criticism you should:

1 **remain quiet and listen**
2 **not find fault with the criticising person**
3 **not manipulate the appraiser by your response (eg despair)**
4 **not try to change the subject**
5 **not caricature the complaint**

6 not ascribe an ulterior motive to the appraiser

7 give the impression you understand the point.

In handling criticism you should accept it and not ignore, deny or deflect it.

Managing meetings

Meetings are much maligned, but are they usually approached and handled as they should be?

In general terms any meeting needs:

- planning
- informality
- participation
- purpose
- leadership

if it is to work, and that is so whether the meeting is in committee or conference format.

A meeting must have a purpose and this can be one (or all) of the following:

- to pool available information
- to make decisions
- to let off steam/tension
- to change attitudes
- to instruct/teach.

Meetings must be prepared for:

1 Know in advance what information, reports, agenda, layout, technical data or equipment is required

2 Be clear about the purpose

3 Inform other participants of the purpose and share, in advance, relevant information/documents

4 Have a timetable and agenda (and notify others in advance)

5 Identify main topics with each having an objective

6 Make necessary housekeeping arrangements.

Chairing a meeting means that you should guide and control it having defined the purpose of it, gatekeeping the discussions as appropriate (opening it to some, closing it when necessary), summarising, interpreting and concluding it with agreed decisions, on time.

The chairman's role in leading/refereeing effective meetings is to ensure that the following elements are handled correctly:

1 *Aim* – after starting on time, to outline purpose clearly

2 *Plan* – to prepare the agenda (and allocate time)

3 *Guide* – to ensure effective discussion

4 *Crystallise* – to establish conclusions

5 *Act* – to gain acceptance and commitment and then to end on time.

Meetings are groupings of people and can develop their own personality. It can help to understand the personality of a particular grouping by reference to group:

- conformity
- values
- attitude to change

- prejudice
- power.

So that the method of running the meeting and making it effective depends on understanding and overcoming problems posed by the group personality.

Within your organisation

Organisations have a degree of permanence, hierarchy and formal communication. Informal communication supplements the formal communication that is needed in organisations.

The **content** of communication in organisations should be (in relation to):

1 **The task:**
   - the purpose, aims and objectives
   - plans
   - progress and prospects

2 **The team:**
   - changes in structure and deployment
   - ways to improve team work
   - ethos and values

3 **The individual:**
   - pay and conditions
   - safety, health and welfare
   - education and training

The **direction** of flows of communication within an organisation must be downward, upward and sideways.

Decisions on what to communicate should bear in mind the must-know priorities and distinguish them from the should-know or could-know lower priorities. The best method for must-know items is face-to-face backed by the written word.

Two-way communication should be used and encouraged to:

- Communicate plans/changes/progress/prospects
- Give employees the opportunity to change/improve management decisions (before they are made)
- Use the experience and ideas of employees to the full
- Understand the other side's point of view.

Summary and six-month follow-up test

Summary

i) Personal reminders

Effective speaking – six key principles:

1 Be clear
2 Be prepared
3 Be simple
4 Be vivid
5 Be natural
6 Be concise.

Practical presentation skills require you to:

- profile the occasion, audience and location
- plan and write the presentation
- use visual aids (if appropriate)
- prepare your talk
- rehearse (with others if necessary)
- deliver on the day.

Good communicators are skilled at listening by:

- being willing to listen
- hearing the message
- interpreting the meaning
- evaluating carefully
- responding appropriately.

Effective writing has three elements:

1 Structure
2 Layout
3 Style

and also needs the six key principles of:

1 Clarity
2 Planning
3 Preparation
4 Simplicity
5 Vividness
6 Naturalness
7 Conciseness.

Six-month follow-up test

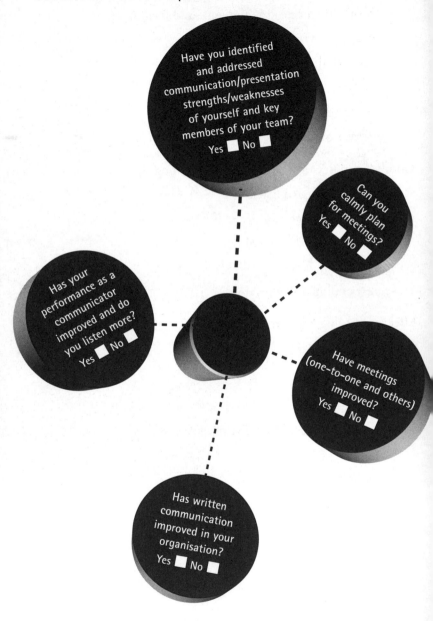

Have you identified and addressed communication/presentation strengths/weaknesses of yourself and key members of your team?
Yes ☐ No ☐

Can you calmly plan for meetings?
Yes ☐ No ☐

Has your performance as a communicator improved and do you listen more?
Yes ☐ No ☐

Have meetings (one-to-one and others) improved?
Yes ☐ No ☐

Has written communication improved in your organisation?
Yes ☐ No ☐

09

PART 2:
PERSONAL REMINDERS
AND THOUGHTS WORTH
THINKING

Introduction

This section (as Chapter 5 in Part 1 did for the first part of the book) will look at the main areas covered in the second part of the book and will offer:

i) personal reminders – points to bear in mind in pursuing your personal development as a leader, motivator and communicator; and

ii) thoughts worth thinking – quotes from various sources which shed light on management and leadership topics covered in Part 2, which you might find helpful or inspiring.

Leadership and teambuilding

i) Personal reminders • Leadership

Whether in team, operational or organisational leadership, what matters is:

- The leader – qualities of personality and character
- The situation – partly constant, partly varying
- The team – the followers: their needs and values

Three overlapping and interacting circles of needs have to be focused on at all times:

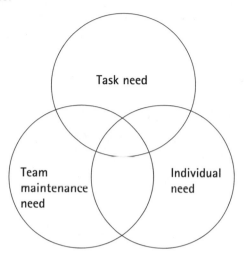

Leadership functions (and the skills needed to exercise those skills) can be summarised as:

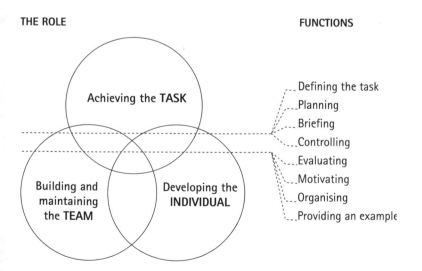

THE ROLE

FUNCTIONS

Achieving the **TASK**

Building and maintaining the **TEAM**

Developing the **INDIVIDUAL**

Defining the task
Planning
Briefing
Controlling
Evaluating
Motivating
Organising
Providing an example

ii) Personal reminders • Teambuilding

Leaders should consider these issues in teambuilding:

Achieving the task

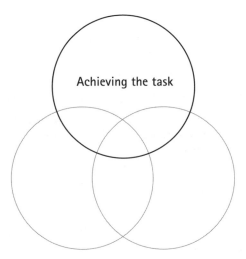

- Purpose
- Responsibilities
- Objectives
- Programme
- Working conditions
- Resources
- Targets
- Authority
- Training
- Priorities
- Progress
- Supervision
- Setting an example.

Building and maintaining the team

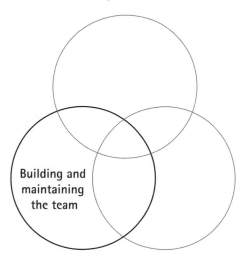

- Objectives
- Standards
- Safety standards
- Size of team
- Team members
- Team spirit
- Discipline
- Grievances
- Consultation
- Briefing
- Representing
- Supporting.

Developing the individual

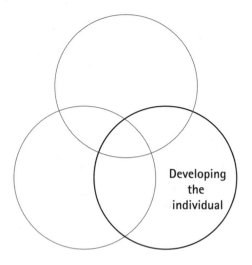

- Targets
- Induction
- Achievement
- Responsibilities
- Authority
- Training
- Recognition
- Growth
- Performance
- Reward
- The task
- The person
- Time/attention
- Grievances
- Security
- Appraisal.

> *'Courage is the quality*
> *that guarantees all others.'*
>
> Churchill

'Leadership is done
from in front.'
Peter Young

'As to moral courage, I have rarely met with the two-o'-clock-in-the-morning kind: I mean unprepared courage, that which is necessary on an unexpected occasion; and which, in spite of the most unforeseen events, leaves full freedom of judgement and decision.'
Napoleon

'President Roosevelt possessed personality, but as his nation's leader... he also did his homework – thoroughly.'

'A sense of humility is a quality I have observed in every leader I have deeply admired.'
Dwight D Eisenhower

Making Time to Think

'What advice can be offered to a leader? He must discipline himself and lead a carefully regulated and ordered life. He must allow a certain amount of time for quiet thought and reflection; the best times are in the early morning, and in the evening. The quality, good or bad, of any action which is to be taken will vary directly with the time spent in thinking; against this, he must not be rigid; his decisions and plans must be adaptable to changing situations. A certain ruthlessness is essential, particularly with inefficiency and also with those who would waste his time. People will accept this, provided the leader is ruthless with himself...

Most leaders will find there is so much to do and so little time to do it; that was my experience in the military sphere. My answer to that is not to worry; what is needed is a quiet contemplation of all aspects of the problem, followed by a decision — and it is fatal to worry afterwards.'

Field Marshal Viscount Montgomery

'Is there not a difference between good leaders and leaders for good?'

John Lord

"A leader is best

When people barely know that he exists.

Not so good when people obey and acclaim him

Worst when they despise him

'Fail to honour people,

they fail to honour you,'

But of a good leader, who talks little,

When his work is done, his aim fulfilled,

They will all say, 'we did this ourselves'."

Loa Tzu • 6th century BC

'Quiet calm deliberation disentangles every knot.'

Harold MacMillan

'Leaders should not be easily provoked.'

St Paul

'The test of leadership is not to put greatness into humanity but to elicit it, for the greatness is there already.'

John Buchan

'He that gives good advice builds with one hand. He that gives good counsel and example builds with both. But he that gives good admonition and bad example, builds with one hand and pulls down with the other.'

Francis Bacon

'It is almost true to say that leaders are 'made' – rather than born.'

Field Marshal Viscount Montgomery

'A good leader must be tough enough to win a fight, but not tough enough to kick a man when he is down.'

W G Bennis and E H Schein

THE PRAYER OF A FAMOUS LEADER

'Lord, make me an instrument of your peace!

Where there is hatred, let me sow love,

Where there is injury, pardon;

Where there is doubt, faith;

Where there is despair, hope;

Where there is darkness, light;

Where there is sadness, joy.

O Divine Master, grant that I may not so much seek

to be consoled, as to console;

to be understood, as to understand;

to be loved, as to love.

For it is in giving that we receive;

It is in pardoning that we are pardoned;

It is in dying that we are born to eternal life.'   Francis of Assisi

'In managing human affairs, there is no better rule than self. restraint.'

Lao Tzu • 6th Century BC

'If you can keep your head when those about you are losing theirs and blaming it on you.'

Kipling

'Your position never gives you the right to command. It only imposes on you the duty of so living your life that others can receive your orders without being humiliated.'

Dag Hammarskjold

ii) Thoughts worth thinking • Teambuilding

'You do not know me, I do not know you, but we have got to work together. Therefore, we must understand each other, we must have confidence in each other. I have only been here a few hours, but from what I have seen and heard since I arrived, I am prepared to say here and now that I have confidence in you. We will work together as a team. I believe that one of the first duties is to create what I call atmosphere. I do not like the general atmosphere I find here — it is an atmosphere of doubt, of looking back. All that must cease. I want to impress upon everyone that the bad times are over and it will be done. If anybody here thinks it cannot be done, let him go at once. I do not want any doubters. It can be done and it will be done beyond any possibility of doubt.'

Field Marshal Viscount Montgomery

Extract from speech to staff in taking over the Eighth Army before the Battle of El Alamein

MORALE

Morale

> Shows itself
> As a state of mind
> Radiating confidence
> In people

Where each member

> Feels sure of his own niche,
> Stands on his own abilities
> And works out his own solutions
> – Knowing he is
> Part of a team

Where no person

> Feels anxiety or fear
> Or pressure to be better
> Than someone else

Where there exists

> A sharing of ideas
> A freedom to plan
> A sureness of worth,
> And a knowledge
> That help is available
> For the asking

To the end that

> People grow and mature
> Warmed by a friendly climate   Anon

'Light
is the
task,
when
many
share
the
toil.'

Homer

Motivation and people management

i) Personal reminders

To draw the best out of people the key strategies are:

- Be motivated yourself
- Select people who are already motivated
- Set challenging but realistic targets
- Remember that progress motivates
- Treat each person as an individual
- Provide fair rewards
- Give recognition

ii) Thoughts worth thinking

'The two great movers of the human mind are the desire of good and the fear of evil.'

Samuel Johnson

'I am persuaded that every being

has a part to play on earth:

to be exact, his or her own part

which resembles no other.' André Gide

**'Such is the state of life that none are happy
but by the anticipation of change.'**

Samuel Johnson

'A man has one eye on what he gives, but seven

eyes on what he receives.' Old German Proverb

'A man's reach should exceed his grasp.'

Robert Browning

'If you treat people as they are, they will stay as they are. But if you treat them as they ought to be, they will become bigger and better persons.'

Goethe

'Give me a fire and I will give you light'

Old Arab Proverb

'Nothing great was ever achieved without enthusiasm.'

Emerson

'No man will find the best way to do a thing
unless he loves to do that thing.'   Old Japanese Proverb

'It is not enough to do our best. Sometimes we have to do what is required.'

Churchill

'Management, above everything else, is about people. It is about the accomplishment of ends and aims by the efforts of groups of people working together. The people and their individual hopes and skills are the greatest variable and the most important one.'

Sir John Harvey-Jones

'You get more of the behaviour you reward. You don't get what you hope for, ask for, wish for, or beg for. You get what you reward.'

Michel le Boeuf

'Fame is the spur that the clear spirit doth raise...
To scorn delights and live laborious days.'

Milton

'Any of us will put out more and better ideas if our efforts are fully appreciated.'

Alexander F Osborn

'No man does anything from a single motive.'

Samuel Taylor Coleridge

'BY ASKING THE
IMPOSSIBLE WE OBTAIN
THE BEST POSSIBLE.'

Italian proverb

'All

that we do

is done

with an eye

to something else.'

Aristotle

Communication and presentation

i) Personal reminders

Effective speaking – six key principles:

1 Be clear
2 Be prepared
3 Be simple
4 Be vivid
5 Be natural
6 Be concise.

Practical presentation skills require you to:

- profile the occasion, audience and location
- plan and write the presentation
- use visual aids (if appropriate)
- prepare your talk
- rehearse (with others if necessary)
- deliver on the day.

Good communicators are skilled at listening by:

- being willing to listen
- hearing the message
- interpreting the meaning
- evaluating carefully
- responding appropriately.

Effective writing has three elements:

1 Structure
2 Layout
3 Style

and also needs the six key principles of:

1 Clarity
2 Planning
3 Preparation
4 Simplicity
5 Vividness
6 Naturalness
7 Conciseness.

ii) Thoughts worth thinking

'Speak properly, and in as few words as you can, but always plainly; for the end of speech is not ostentation, but to be understood.' William Penn

'Communication is the art of being

UNDERSTOOD.'

Peter Ustinov

'What is conceived well is expressed clearly
And words to say it will arise with ease.'

Nicholas Boileau

'If any man wishes to write in a clear style, let him first be clear in his thoughts.'

Goethe

'Have something to say
and say it as clearly as you can.
That is the only secret of style.'

Matthew Arnold

'**Reading**

is to the **mind**

what exercise

is to the **body**.'

English Proverb

'The major mistake in communication is to believe that it happens.' George Bernard Shaw

'In good communication, people should be in no doubt that you have reached a **conclusion**.'

John Adair and Neil Thomas